Stress and Strain Engineering at Nanoscale in Semiconductor Devices

Stress and Strain Engineering at Nanoscale in Semiconductor Devices

Chinmay K. Maiti

CRC Press
Taylor & Francis Group
Boca Raton London New York

CRC Press is an imprint of the
Taylor & Francis Group, an **informa** business

First edition published 2021
by CRC Press
6000 Broken Sound Parkway NW, Suite 300, Boca Raton, FL 33487-2742

and by CRC Press
2 Park Square, Milton Park, Abingdon, Oxon, OX14 4RN

© 2021 Taylor & Francis Group, LLC

CRC Press is an imprint of Taylor & Francis Group, LLC

ISBN: 978-0-367-51929-2 (hbk)
ISBN: 978-0-367-51933-9 (pbk)
ISBN: 978-1-003-05572-3 (ebk)

Typeset in Times
by SPi Global, India

All my teachers from day one ….
And
All my students from day one, I started teaching …

Contents

Preface

In the era of the Internet of Things, the road toward enhancing the performance of integrated circuits comprises the fundamental improvement of the devices via new structures, scaling the size and transistor features, which increase their speed and reduce power consumption. Anticipating a limit to this continuous miniaturization (More-Moore), intense research efforts are being invested to co-integrate various functionalities (More-than-Moore). In this context, flexible and stretchable electronics offer many opportunities. The combination of new device architectures and the use of stress/strain tuning and high-mobility materials have led toward nanoelectronic devices and pushed the limits of CMOS. Computer-aided design and simulation of nanoscale transistors for upcoming technology nodes (7 nm and below) are the main focus of this book. The monograph addresses mainly the design issues of advanced nanodevices with technical depth and conceptual clarity and presents leading-edge device design solutions for 7 nm technology nodes.

The purpose of this monograph is to bring into one resource presenting a comprehensive perspective of advanced nanodevices design using advanced simulation tools. Process simulation is similar to the virtual manufacture of semiconductor devices. As per IRDS, 3D integration has become one of the main directions to fit into device performance roadmaps. After a brief discussion on basic device physics, the readers are introduced to device design and simulation. This monograph presents the design, simulation, and analysis of Si, heterostructure silicon-germanium (SiGe), and III-N compound semiconductor devices. After the introduction, the monograph presents seven Chapters viz., the Simulation Environment, Stress Generation Techniques in CMOS Technology, Electronic Properties of Engineered Substrates, Bulk-Si FinFETs, Strain-Engineered FinFETs at NanoScale, Technology CAD of III-Nitride Based Devices, and Strain-Engineered SiGe Channel TFT for Flexible Electronics.

It is the purpose of this book to give device engineers, who are new to the field, a quick start on device and technology design using the TCAD tools. Detailed and extensive technology CAD simulations have been presented for stress- and strain engineering in 3D nanodevices. This monograph attempts to fill the gap in the literature in a rapidly evolving field as it blends a wide-ranging description of TCAD activities in the process to device design in Si, SiGe, and III-N materials, technology, and their applications. The monograph is primarily intended for senior undergraduate and graduate students and professors who wish to find a technology CAD teaching reference book and others who are interested in learning about semiconductor device design using simulation. This monograph may be used as a reference for engineers involved in advanced device and process design. Approaches described in this monograph are expected to boost advanced device design using challenging TCAD simulations and help with characterizing new processes and devices.

I am especially grateful to all my research students whose association has been the source of learning for many new topics covered in this monograph. Special thanks go to Drs. T. P. Dash and S. Das, who have kindly contributed three chapters (Chapters 3, 5, and 6). We would also like to express our deep appreciation to the CRC team for their helps during the preparation of the book. Finally, I would like to thank my family members (my wife Bhaswati, sons Ananda and Anindya) for their support, patience, and understanding during the preparation of the manuscript.

C. K. Maiti
Kolkata
December 2020

Author Biography

Prof. Chinmay K. Maiti, PhD, is an Ex-Professor and Ex-Head of the Department of Electronics and ECE in Indian Institute of Technology (IIT) – Kharagpur, India. He then joined the SOA University, Bhubaneswar, in May 2015 as a professor, where he is now on a visiting assignment. He is interested in strain engineering in nanodevices, flexible electronics, and semiconductor device/process simulation research, and microelectronics education. He has published several monographs in Silicon–Germanium, heterostructure-Silicon, and Technology CAD areas. He was the editor of the *Selected Works of Professor Herbert Kroemer*, World Scientific, Singapore, 2008.

1 Introduction

The semiconductor industry has come a long way since its inception with the first transistor in 1947. This industry saw the miniaturization days of Dennard, during which everyone was happy to reduce the vertical and horizontal dimensions of the devices and see performance improvement [1]. Moore's law [2], which began as an observation of Gordon Moore, co-founder of Intel, has been accepted as law by the industry with everyone trying to keep the law alive by doubling the number of transistors every 2 years. At the end of the last century, the industry began to face difficulties with miniaturization devices, which were then solved using new materials like high-permittivity gate dielectrics or metal gate stacks [3]. Now, once again, the industry soon expects to hit the physical boundary wall of miniaturization of silicon, which has seen the birth of new innovative devices such as Fully Depleted Silicon-On-Insulator (FDSOI) [4] and FinFET [5], and 3D integration techniques [6], making it possible to rethink computer architectures in terms of artificial intelligence and quantum computing [7]. More-Moore scaling involves: 1) Device Architecture, 2) Performance Boosters, 3) FDSOI, 4) Lateral Gate-All-Around device (LGAA), 5) Vertical GAA, and 6) Tunneling FET. Miniaturization is symbolized by Moore's law, empirically stating that the number of transistors in a processor doubles about every 2 years while the price per component will be halved. This law, first proposed in 1965 and then refined in 1975 [8] by Gordon Moore, has been verified in practice until the end of the last century. It was considered as an objective by the International Technology Roadmap for Semiconductors (ITRS). Since the early 2000s, a simple reduction of dimensions is no longer sufficient to verify Moore's law [9].

1.1 BEYOND SILICON

New processes, architectures, and materials have emerged to continue the quest for doubling computing power every 2 years. It's about the More-Moore era. Previously managed by the ITRS, the specification roadmaps logic circuits have been fixed since 2016 by the International Roadmap for Devices and Semiconductors (IRDS). Table 1.1 is an extract from the specifications published by IRDS in 2017. We see in particular the structure of the components, their dimensions (length and gate width) as well as the supply voltage. An alternative to Moore's dedicated silicon strategy, the More-than-Moore (MtM) strategy, aims to improve the performance of MOSFETs using other materials and creating compact systems and systems on chips. In addition to reducing the dimensions of the components, the successive recommendations of the ITRS and then the IRDS show a decrease in supply voltages. For digital circuits, this implies that the intrinsic noise (generated by the component itself) could eventually become a constraint because too high a noise level can distort logical levels. At a lower supply voltage, a component will therefore be

1

TABLE 1.1

Extract from the IRDS Roadmap for the Specification of Logical Components [10]. (L-GAA: Lateral Gate-All-Around; V-GAA: Vertical Gate-All-Around)

Year of Production	2017	2019	2021	2024	2027	2030	2033
Technology node (nm)	10	7	5	3	2	1.5	1.0
Component structure	FinFET	FinFET	L-GAA	L-GAA	L-GAA	V-GAA	V-GAA
Supply voltage (V)	0.75	0.75	0.70	0.65	0.65	0.60	0.55
Gate length (nm)	20	18	16	14	12	12	12
Minimum width (nm)	8.0	7.0	7.0	7.0	6.0	6.0	6.0

more subject to logic level errors, especially as the intrinsic noise level increases when the dimensions decrease.

Advances made during the More-Moore era [11, 12] have led to a lot of problems to be solved, such as short channel effects, the complexity of technological processes, and their performance, the reliability of the devices, and their variability. The increase in the number of transistors has made it necessary to keep the constant frequency clock to limit the power consumed by the technology. Clocking frequency at nominal supply voltage is expected to be improved from 2.5 GHz in 2017 to 4.2 GHz in 2033. Further miniaturization for low-power devices and new applications (like the Internet of Things [IoT]) require more power reduction of the devices, which is only possible with a decrease in the supply voltage.

To maintain the scaling at low voltages, scaling in recent years focused on additional techniques to boost the performance, such as the use of introducing strain to channel, stress boosters, high-k metal gate, lowering contact resistance, and improving electrostatics. This was all done to compensate for the gate drive loss, while supply voltage needs to be scaled down for high-performance mobile applications [13, 14]. One of the main challenges to achieving desired performance depends on the variability of the devices because the decrease in the supply voltage also reduces the margin available for circuit design due to variability [15]. System scaling enabled by Moore's scaling is increasingly challenged by the scarcity of resources such as power. The following applications drive the requirements of More-Moore technologies that are addressed in the IRDS [7]. The More-Moore roadmap focuses on sustaining the performance scaling at scaled dimensions and scaled supply voltage. It is expected that in 2024, p- and n-MOSFETs could be stacked on top of each other, allowing a further reduction (see Figure 1.1). Figure 1.2 shows the background and motivation behind switching to 3D devices [16].

Table 1.2 shows the technology roadmap for nanowire and FinFETs. After 2027, there is no room for 2D geometry scaling where 3D will be needed for very-large-scale integration of circuits and systems using sequential/stacked integration approaches [17]. This is because there is no room for contact placement as well as worsening performance as a result of the contacted poly pitch (CPP) scaling and metal pitch scaling.

It is anticipated that bulk silicon will remain the mainstream substrate. Finding solutions to these issues has slowed down the shrinking of the device size, leading to

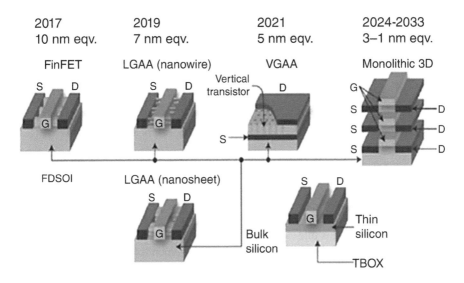

FIGURE 1.1   IRDS future generation device architecture roadmap [10].

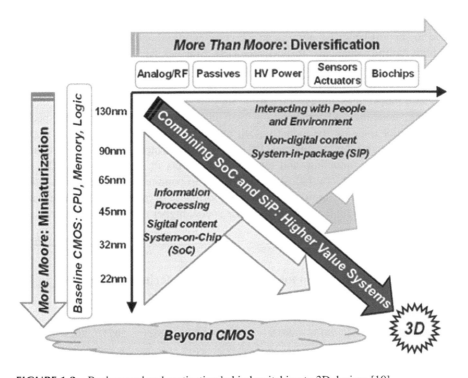

FIGURE 1.2   Background and motivation behind switching to 3D devices [10].

TABLE 1.2

IRDS Technology Roadmap for Nanowire and FinFETs [10]

Key Specifications	FinFET			Nanowire		
	5 nm Node	4 nm Node	3 nm Node	5 nm Node	4 nm Node	3 nm Node
Channel length (nm)	15	13	11	15	13	11
Fin/wire total width (nm)	7.5	6.5	5.5	15	13	11
Fin/wire total width (nm)	30	30	30	6	5.2	4.5
Cladding thickness (nm)	2	1.75	1.5	1.5	1.25	1
Fin/wire pitch (nm)	30	26	22	30	26	22
Access resistance (ohm)	808	938	11,145	2,314	2,879	3,708

the use of new materials and device designs. The semiconductor industry has been able to reinvent itself keeping a continuous improvement of the performance of its product. There is still a long way to go before the MOSFET technology will be exhausted. Meanwhile, the semiconductor industry has been able to find solutions to these issues, leading to the use of new materials and device designs. In consequence, new transistor architectures are considered to replace standard technology [18]. The performance demanded by IRDS for future technology nodes is shown in Table 1.3. At the same time, silicon-on-insulator (SOI) and strain relaxed buffer (SRB) will be used to support better isolation (e.g., RF co-integration) and defect-free integration of high-mobility channels, respectively [19]. SOI has the advantage of threshold voltage tuning, allowing to tune a device to either high-performance or low-leakage to get back better gate control.

These innovations are expected to be introduced at a rapid pace. Hence, understanding, modeling, and implementing them into manufacturing on time are expected to be a major issue for the industry [20]. SiGe and Ge channels are gaining importance as the high-mobility channels [21]. III-V channel faces challenges of variability, band-to-band tunneling, and massive investments in manufacturing [22]. The goal of the semiconductor industry is to be able to continue to scale the technology in overall performance improvement. The performance of the final chip can be evaluated in many different ways such as higher speed, higher density, lower power, and

TABLE 1.3

IRDS Technology Roadmap for Actual and Future Technology Nodes Predicted for High-Performance Devices, Where L_G is the Physical GateLength and V_{DD} is the Supply Voltage [10]

Year	2013	2015	2017	2019	2021	2023	2025	2027
Technology Node (nm)	14	10	7	5	3.5	2.5	1.8	1.3
Metal half-pitch (nm)	40	32	25.3	20	15.9	12.6	10	8
L_G (nm)	20.2	16.8	14	11.7	9.7	8.1	6.7	5.6
V_{DD} (V)	0.86	0.83	0.80	0.77	0.74	0.71	0.68	0.64

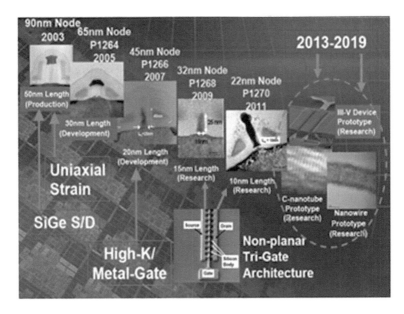

FIGURE 1.3 Summary of currently available microelectronic devices in manufacturing [10].

more functionality. A summary of currently available microelectronic devices in manufacturing is shown in Figure 1.3.

As the gate length shrank even further, at around 30 nm, short channel effects became very serious requiring paradigm shifts in the MOSFET architecture to continue scaling. Effective gate control on the channel had to be increased considerably to reduce detrimental short channel effects below 30 nm. It has been demonstrated that this could be achieved reliably through multigate FET architecture or fabrication of planar devices on ultra-thin SOI substrate. The evolution of multigate transistor structures is shown in Figure 1.4. SOI and multigate technologies have been proposed to improve electrostatic control by the gate and strongly reduce the SCE. The first commercial multigate MOSFET was released in 2011 by Intel using the structure of a bulk FinFET.

IRDS provides specifications typically for two types of applications: one for low-power technology (mobile, laptop), and high performance for logic applications

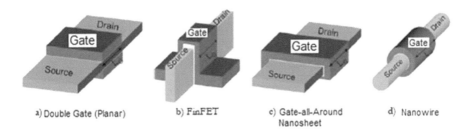

FIGURE 1.4 Evolution of multigate transistor structures.

(microprocessors). High-performance applications require an increase in the switching speed for logic, which is directly related to the current supplied by the transistor (I_{on}), and a reduced leakage current (I_{off}). The best architecture for a transistor is where we get the best compromise I_{on}/I_{off} ratio with good control of short channel effects. For the next device generation, multigate architecture with an extremely shrunk body (nanowire) on bulk or SOI substrates has been proposed. The gate length of MOSFET should become sub-10 nm (5 nm) in 2026.

1.2 TOWARD NEW TRANSISTOR ARCHITECTURES

Although numerous, the technological innovations made using conventional architecture on bulk Si substrates, are no longer sufficient to meet the performance specifications (I_{on}/I_{off} and I_{eff}/I_{off}), mainly because of the electrostatic control which is becoming more and more difficult to obtain. The introduction of new architectures, and therefore the end of CMOS platforms based on conventional architecture on a solid substrate is approaching fast. Figure 1.5 shows the ITRS predictions given by the 2011 edition of its roadmap for low operating power and low supply voltage. These predictions are getting confirmed, within a few years, by the announcement and various publications, for example, the announcement of the CMOS FDSOI platform for the 28 nm node [23]. Intel announced its 22 nm platform using the trigate architecture [24] already marketed as high performance (microprocessors), and the use of the architecture is intended for mobile applications or System on Chip (SOC) applications.

Even though it is accepted that Moore's law has slowed down, the CMOS technology scaling has however not stopped yet, helped by design/technology co-optimization (DTCO). New architectures have replaced the historical bulk planar devices. FinFET and FDSOI technologies have demonstrated excellent electrostatic control. Strain engineering remains one of the most powerful tools to increase the performance of CMOS technology [26]. It is indeed today discussed for achieving sub-10 nm node requirements [27]. Finally, other essential factors need to be considered when developing technology in an industrial context such instance the variability [28], the reliability [29], or the yield.

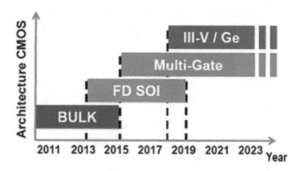

FIGURE 1.5 Prediction of the use of different CMOS architectures by ITRS according to the 2011 roadmap for low-power devices [25].

1.3 FUTURE OF NANO-CMOS TECHNOLOGY

To extend CMOS scaling beyond the conventional bulk device limit of about 10 nm gate length, intense research activities are focused on alternative materials and device structures including high-k gate dielectrics, high-mobility strained silicon channel, ultra-thin SOI, and double-gate MOSFETs. Some of the alternatives being actively pursued for future technology nodes are the multigate device architectures, alternate channel materials with multigate FETs (III-V and strained-Ge quantum wells, and hetero-FETs), super-steep sub-threshold slope transistors (sub-60 mV/dec), graphene-based FETs, and spin FETs. All of these approaches promise low-power operation.

Strain engineering has been the most effective for performance enhancement in the last decade, as illustrated for the 32 nm node and earlier [30]. However, the effects of those stressors may not extrapolate intuitively into newer nodes. With the scaling down of CPP, SiGe on the source/drain epitaxial (S/D EPI) contact and SRB remain effective boosters to scale mobility more than double on top of high-mobility channel material [31]. SiGe channel for p-MOSFET and strained Si channel for n-MOSFET has been successfully demonstrated on a 7 nm CMOS platform using SRB [32]. SRB or S/D stressors may not be useful for channel stress generation in vertical devices, which appear in the roadmap around 2021.

Other strain engineering techniques also contain gate stressor and ground plane stressors, which adopt the beneficiary vertical stress components for NMOS. Compressively strained SiGe channel is also shown in ultra-thin body and buried oxide (BOX) fully depleted SOI (UTBB FDSOI) to boost pFET performance [33, 34]. A high level of stress is maintained in the channel thanks to the planar configuration (with a low aspect ratio, compared to FinFET). Combined with the use of back-bias (to reduce V_{DD} and thus the dynamic power), it enables high-performance, low-power circuits on UTBB FDSOI.

Parasitic capacitance between gate and source/drain terminal of the device is expected to increase with technology scaling. This component is getting more important than channel capacitance-related loading whenever the standard cell context is considered and elevated in the GAA structures as a result of unused space between consecutive devices. There is a need to focus on low-κ spacer materials and even air spacers that still provide good reliability and etch selectivity for S/D contact formation [35, 36]. It appears that there are significant limits in increasing FinFET or lateral GAA device AC performance by increasing the height of the device.

FinFET has better electrostatics integrity due to its tall narrow channel that is controlled by a gate from three sides that allow relaxing the scaling requirements of fin thickness (i.e., body thickness) compared to UTBB FDSOI. In UTBB FDSOI, electrostatic control could be established by using silicon (i.e., body) thickness and BOX thickness where convergent scaling of both silicon thickness and BOX thickness enables electrostatics scaling, that is, drain-induced barrier lowering (DIBL) below 100 mV/V down to gate length beyond 10 nm [37]. Besides the channel leakage induced by electrostatics, there are potentially other leakage sources such as sub-fin leakage or punch-through current. This leakage current flows through the bottom part of the fin from source to drain. This gets more problematic in Ge channels because of the low effective mass of Ge. Ground plane doping and quantum well below the channel would potentially solve this leakage problem, therefore improving the electrostatics [38]. Reducing variability would further allow V_{DD} scaling. Controlling the channel length and the channel thickness is important to maintain the electrostatics in the channel. This would require, for example, controlling the profile of the fin and lithography processes to reduce the critical dimension uniformity (CDU), line width roughness (LWR), and line edge roughness (LER). Dopant-free channel and low-variability work function metals would reduce the variations in the threshold voltage. With the introduction of high-mobility materials, gate stack passivation is needed to reduce the interface-related variations as well as maintaining the electrostatics and mobility.

But the race for miniaturization now faces very serious difficulties. Indeed, difficulties related not only to physical phenomena but also the realization technologically, appear at the level of transistors when one arrives at a subnanometric channel depending on the trend called "More-Moore." Fortunately, to increase chip performance, miniaturization is not the way to go but the only way out. It can be bypassed, or even associated, by another path of integration, viz., the integration of heterogeneous chip, called "More than Moore." It is recognized as a powerful solution allowing heterogeneous co-integration while increasing density and performance in terms of consumption and bandwidth. Offering new degrees of freedom, this technology is arousing the enthusiasm of many players in the domain as evidenced by the literature, projects, and applications that emerge. This evolution is made possible in particular by the mastery of through-silicon-via (TSV) technology – connections that pass vertically through the silicon and make it possible to connect the chips with different functionality assembled on top of each other (see Figure 1.6).

It became possible a few years ago to jointly manufacture electronic circuits and sensory components, such as sensors and actuators while having the possibility of co-integrating them; in particular, thanks to the growing compatibilities of manufacturing processes [40]. However, the problems of integration density and therefore of connectivity are coming back to the fore. To get out of this impasse, the concept of 3D integration of chips emerged. The concept of 3D integration constitutes a paradigm shift because the architecture of the chip needs to be almost completely redesigned. Indeed, new issues and constraints linked to the design, manufacture, assembly, and packaging appear and constitutional challenges to be taken up, quickly, to guarantee the promised performance. 3D integration is therefore now widely recognized as a solution effective in overcoming the challenges of miniaturization

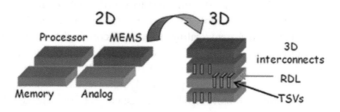

FIGURE 1.6 Concept of 3D integration [39].

and chip densification. It's a concept that combines More-Moore and More-than-Moore [41, 42].

From the point of view of manufacturing, the compatibility between SOI and conventional technology resulted in a rapid consolidation of both structures; hence, it became extensively used in industry [43] as well as in the academy [44]. Some of the indispensable steps in conventional MOSFETs to guarantee the correct isolation between devices, and to reduce the parasite effects are unnecessary in SOI and the lateral dielectric. For this reason, the chips based on SOI are compact and fabricated more easily. Nevertheless, the manufacturing cost of SOI is slightly higher than for conventional devices because wafers have to be pre-processed to get the substrate. Another problem of this technology is the influence of Si film thickness on the electron mobility degradation due to phonon confinement, Coulomb scattering, and doping fluctuations [45, 46]. FDSOI technology is fully compatible with existing manufacturing facilities. It can significantly reduce the number of steps involved in the device fabrication process due to the use of an insulator substrate. Both planar and 3D devices can be manufactured. 3D devices can be integrated by stacking packages, die stacking, or TSV technology [47]. Sustaining increased transistor densities following Moore's law has become increasingly challenging with limited power budgets and fabrication capabilities [48].

FinFETs could potentially take us to the 5 nm node, but what comes after it? From gate-all-around devices to single-electron transistors and 2D semiconductors, a torrent of research is being carried out to design the next transistor generation, engineer the optimal materials, improve the fabrication technology, and properly model future devices. Insights from investigators and scientists in the field have led to research papers, short communications, and review articles that focus on trends in micro- and nanotechnology from fundamental research to applications.

1.4 TECHNOLOGY CAD

In a highly competitive semiconductor product environment, the advantages of the use of the predictive power of Technology Computer-Aided Design (TCAD) are now well established to reduce cost and time in product development [49]. Through proper calibration with prototype wafers, TCAD can also accurately predict the behavior in the manufacturing line for new technology nodes. This information, coupled with design tools, enables the manufacturing line to fine-tune the process for specific designs and, ultimately, to achieve higher yields. From a technology and modeling point of view, a combination of process and device simulation is at the core of TCAD, as it bridges between a given technology and circuit design. TCAD is now playing an increasingly important role in the virtual wafer fabrication (VWF) approach to design and technology development. Among various possible approaches for analyzing microelectronic systems, the most common approach may be classified into three levels: device, circuit, and systems.

To understand the challenges facing advanced device design, one must first understand the design parameters available to a device designer and their significance. The goal of device design is to obtain devices with high performance, low power consumption, low cost, and high reliability. Instead of going through an

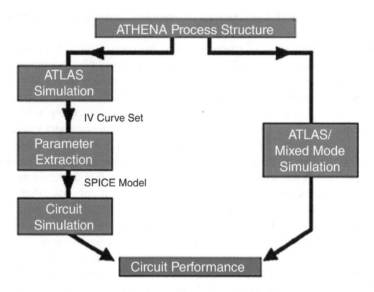

FIGURE 1.7 Flow diagram for integrated TCAD-based device design viz., the process of device simulation and SPICE model parameter extraction essential for circuit design (Source: Silvaco).

expensive and time-consuming fabrication process, computer simulations can be used to predict the electrical characteristics of a device design quickly and cheaply. TCAD can also be used for reducing design costs, improving device design productivity, and obtaining better device and technology designs (see Figure 1.7). TCAD consists mainly of two parts: process design, and device simulation. TCAD is now an indispensable part of semiconductor modeling and design. In the semiconductor industry, the core production process of manufacturing integrated circuits (ICs) takes place in a front-end manufacturing facility. The wafers go through a specific processing sequence that consists of a certain number of processing steps. Modern complex semiconductor manufacturing processes consist of a large number of processing steps, long processing time, dynamic interactions among different tools, and complex interrelations between tool performances and product qualities.

In this monograph, we present the outlook for the virtual manufacturing of advanced semiconductor devices for technology node 10 nm and below. It is particularly intended to address the design and simulation issues associated with the front-end aspects of extending CMOS technology. Areas covered include transistor structures, front-end materials, and front-end processes along with their device processing and simulation, characterization and modeling, and the simulation infrastructure. As the monograph encompasses broad areas of semiconductor technology, physics, modeling, and characterization, etc., and this book is planned as a design book, sources have been referred for interested readers for detail. Simulation-based scaling study – including device design, performance simulation, and the impact of variability – on nanoscale devices is considered.

This book is intended both as a self-learning resource or professional reference and as a text for use in graduate courses. A key feature of the book is modular organization. The seven chapters may be read in almost any order. All simulation results presented throughout this monograph were obtained by the authors using the GTS Framework [50] and Silvaco [51] tools.

1.4.1 CHAPTER 2 SIMULATION ENVIRONMENT

TCAD was originally pioneered at Stanford University. The monograph is associated with the recommendations of the IRDS roadmap [10] concerning the efforts made by industry and academic partners in the modeling of future generations of transistors. To create a foundation for the simulation environment and to ensure reproducibility, the used material models and tools are introduced in this chapter. The evolution of semiconductor simulation tools goes back to the late 1960s. Starting in the late 1960s, the process and device simulation were predominantly in 1D and 2D. Device simulation is dominantly 2D due to the nature of devices and became the workhorse in the design and scaling of devices. Currently available advanced semiconductor device/process simulation tools are introduced. There are several commercial TCAD software platforms for process/device simulation around the world. Most of them were first developed by the universities worldwide and later integrated into a suite by the TCAD vendors. Since the early 1990s, commercial vendors such as SILVACO and SYNOPSYS developed their graphical user interfaces around existing frameworks.

The theoretical modeling of 2D and 3D nanoscale transistors demands a proper treatment of quantum effects such as the energy level quantization caused by strong quantum confinement of electrons and band structure non-parabolicity. Electron-phonon scattering, surface roughness, alloy disorder, and tunneling should be included. To address these issues, a multidimensional quantum transport solver based on a self-consistent solution of the Schrödinger and Poisson equations in the real-space effective mass approximation with a tight-binding extraction of the effective mass values is used to simulate III-V and strained-Si devices. In the nanodevices transport simulation framework, approaches that incorporate important quantum effects into semi-classical models have become very popular due to their lower computational cost in comparison to the purely quantum transport simulation techniques.

TCAD approaches yielding accurate physical insight and useful predictive results for real-world semiconductor applications are described using multidimensional (2D and 3D) simulations. In this chapter, a brief background on past and present simulation development activities has been given. An overview of currently available commercial TCAD tools, viz., GTS Framework, Silvaco, and Synopsys, has been presented. Simulation tools developed in the framework of self-consistent Poisson–Schrödinger for the electrostatics, Quantum Drift-Diffusion (DD), and Mode Space (MS) approaches for solving the carrier transport models, Kubo–Greenwood model to account for scattering have been introduced. State-of-the-art device simulation tools combine: (i) accounting for quantum effects based on the rates of the relevant multi subband scattering mechanisms [52] and (ii) using the semi-classical Boltzmann transport equation (BTE) in the relaxation time approximation by adopting the Kubo–Greenwood formalism [53, 54].

1.4.2   Chapter 3 Stress Generation Techniques in CMOS Technology

Until the 1980s, stress has long been synonymous with defects (e.g., gaps, impurities) [55] or extensive dislocations in microelectronic materials and delamination of layers. It's only since 2000, semiconductor industries have begun to exploit the influence of mechanical stress on mobility to improve the performance of MOSFET transistors [21, 56]. The strain modifies the semiconductor band structure, which implies both a modification of the energy of the forbidden band as well as a lifting of degenerations of the bands (conduction and valence), reducing the probability of interaction between carriers and phonons.

Controlling and understanding the stress in materials is of major importance in the successful fabrication of microelectronic devices. Failure to properly account for stress-related effects can lead to substrate warping and layer delamination, both of which are detrimental to the performance and reliability of components. Hence, it is desirable to have a reliable and automated simulation methodology spatially monitor both stress and strain in nanodevices. This technique will become a powerful means of improving the performance and optimizing stress design for strain-engineered devices at 22 nm node and below. Strain mapping techniques for performance improvement of strain-engineered devices are discussed. In this chapter, we shall describe the method used in the simulation of deformations. Linear elasticity basics will be recalled, particularly in the case of cubic structure materials.

In general, two types of strains can be generated at the time during the manufacture of devices. On the one hand, the unintentional mechanical strain is systematically generated throughout the process. These strains have different origins, such as chemical mechanical polishing, ion implantations, thermal annealing, oxidation, and intrinsic strain that depend on the conditions of the deposition. The second type relies on the strains induced intentionally during the manufacturing process to improve the performance of these devices, such as the use of constrained substrates: sSOI or SiGeOI.

Since the 2000s, MOSFET performance enhancements have been achieved mainly via device engineering while keeping the gate dielectric thickness almost constant. Three major innovations viz., channel stress and strain engineering for mobility enhancement (90 nm and 65 nm nodes), high-k/metal gate technology for reduced gate leakage (45 nm and 32 nm nodes), and trigate device architecture for improved electrostatic control (22 nm node) have been introduced. The summary of various types of technology boosters via stress/strain engineering in front-end processing is shown in Figure 1.8.

1.4.3   Chapter 4 Electronic Properties of Engineered Substrates

In Chapter 3, the theory of stress, strain, and their interdependence has been discussed. In this chapter, we consider in particular the evolution of the piezoresistive effect and carrier transport properties. The impact of strain on the band structure of silicon is detailed. We discuss the effects of process-induced stress on the conduction and valence bands of silicon that will allow us to understand the use of piezoresistivity in MOSFET technologies to improve performance. Piezoresistive coefficients are defined

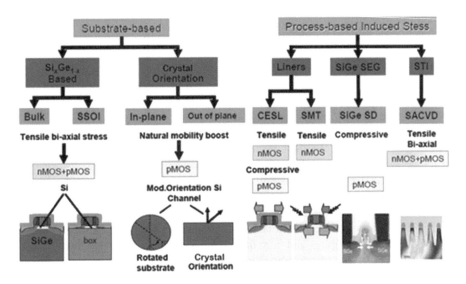

FIGURE 1.8   Schemes of front-end stressors types [20].

to describe the effect of stress on mobility in semiconductor materials. In planar CMOS technologies, orientation-dependent mobility enhancement has been demonstrated through the use of hybrid orientation technology (HOT) [57, 58]. Hole mobility increases significantly when the channel orientation changed from Si (100) to Si (110). However, the electron mobility is severely degraded with the same orientation change. Although the mechanisms behind the enhancement in stress/strain-induced mobility are fairly well understood qualitatively, quantitative evaluation is much more difficult as the type of mechanical stress-induced is indeed very complex. Moreover, the implications of surface orientation on mobility are discussed. A model of the evolution of the piezoresistive coefficients has been presented from a transport model for transistors. This model makes it possible to predict the variations of the piezoresistive coefficients with the cross-section, viz., width, and thickness of the channel.

It is important to know how stresses from different sources interact with each other in areas of interest, for example, under the gates of various devices in an inverter cell. The interaction effects could be even more pronounced when the critical dimensions of individual devices and distances between them scale down. In a simulation case study, we shall use the 3D simulator VictoryStress [59] to analyze the stress effects on carrier mobilities of individual n- and p-FinFET devices. We shall show that a combination of the 3D process simulator VictoryCell and 3D stress simulator VictoryStress allows fast and accurate stress analysis of complex cell structures, such as an inverter cell.

1.4.4   CHAPTER 5 BULK-SI FINFETS

The first part of the chapter will be devoted to the evolution and the presentation of advanced solutions to enable the development of multigate architectures to overcome the limitations that miniaturization imposes [60]. As the gate length is reduced, the difficulties in managing short channel effects in planar bulk MOSFET architectures have led to the favor of FinFET and FDSOI architectures [61]. The electrostatic control of the gate on the channel is excellent and the performance is increased, thanks to the gain in effective width in the case of the FinFET while the FDSOI takes advantage of the threshold voltage modulation effect by the polarization of the substrate. However, after several generations are produced, the question is whether the dimensions of these technologies will be able to continue to be reduced. FinFET remains the key device architecture that could sustain scaling until 2021 for high-performance logic applications [62]. We shall discuss how performance can be improved through different manufacturing methods such as the introduction of mechanical stress. The use of compressive stress and strain shows possible improvements in the performances of the transistors.

Trigate FinFET full virtual fabrication process is presented for a 7 nm technology node. For process optimization, suitable models have been adopted for the simulation of oxidation, lithography, and ion implantation. We shall present results from numerical simulations of the fabrication of TriGate FinFET transistors on bulk Si substrates. We shall examine the strain distribution in FinFET structures with epitaxial $Si_{1-x}Ge_x$ stressors deposited around the source/drain region. The SiGe source–drain stressor has been used to generate compressive stress in the channel. The linear elasticity theory has been implemented in simulation to describe the evolution of the stress for various stages of the process. We have been able to predict the possibility of tuning the residual stress by changing the geometry of the SiGe stressor (shape) during epitaxial growth. We shall demonstrate the role of the stress/strain on device variability in a state-of-the-art trigate FinFET with SiGe source/drain stressors (e-SiGe) at a 7 nm technology node using full 3D TCAD simulations. The stress mapping technique is used to quantify strain distribution in the devices. It is shown that a shallower and closer-to-channel e-SiGe stressor with higher Ge content could increase compressive stress and therefore improve the drive current through hole mobility enhancement.

Toward electrical performance evaluation through simulation, the effects of quantum confinement are taken into account. The Bohm Quantum Potential (BQP) model is used to introduce quantum confinement in 3D. We shall show how using TCAD mechanical stress tuning techniques and electrical performance simulation, one can design advanced FinFETs. Mechanical stress tuning techniques help in significant improvements in the performance of FinFET at the 7 nm technology node.

1.4.5   Chapter 6 Strain-Engineered FinFETs at NanoScale

The e-SiGe technique has emerged to be a consistent performance booster for advanced devices below 14 nm technology node whose process simulation has been discussed extensively in Chapter 5. In this chapter, we use an extensive 3D TCAD simulation framework [49] to demonstrate how the elastic stress models can be applied to the simulation of stress transfer using source/drain SiGe-epi layer in 7 nm technology nodes for p-FinFETs. The omnipresent residual stress is now becoming an important source of variability in advanced VLSI technologies that influence circuit performance. The design and optimization of FinFETs at the nanoscale regime including variability are thus extremely important [63]. In deeply scaled technologies, process and environment variations become important sources of variabilities [64, 65]. The variability sources include roughness of the lines induced by the lithography and etching and the granularity of the metal gate linked to deposition conditions. The gate may be irregular and show grains due to its polycrystalline character.

Metal grain granularity (MGG) and random discrete dopants (RDDs) are the major sources of process-induced variability in FinFETs [66, 67]. MGG is an issue mainly connected with the "gate first" process technology, where the gate metal is deposited before any high-temperature annealing procedure [68]. Depending on their crystal orientations, the work function of each metal grain varies randomly. Therefore, instead of a single work function value, the transistor gate contains multiple grains with different work function values that cause random fluctuation of device performance parameters [69, 70]. It is shown that RDDs in the source–drain critically affect the performance through the variations in the short channel effects. In this chapter, the device critical performance parameters such as threshold voltage, I_{on}, I_{off}, subthreshold slope variation due to RDD, and MGG are examined in detail. The impact of location-dependent discrete dopants in different positions like the source/drain side has been studied to predict electrical performance. We calculate the mean and standard deviation of these parameters (QQ plots) to quantify the variability.

A methodology is proposed to simulate the effects of the local work function variation due to MGG on device matching. Using this methodology, the threshold voltage distribution will be bound within the extreme values set by the work function. Simulation results show that the process-induced variability sources become important for FinFET devices as they affect more the gate control and the channel charge. Stress profiling analysis is adopted to simulate process steps of SiGe epitaxial S/D p-FinFETs. This study will provide a guideline for trigate FinFET design using stress mapping/tuning which is of great importance for technology scaling.

1.4.6 CHAPTER 7 TECHNOLOGY CAD OF III-NITRIDE BASED DEVICES

The rising III-nitride compound semiconductor materials have superior transport properties, high breakdown electric fields or high thermal and/or high heat capacity, and thermal conductivity are suitable for applications that are driven more by performance and less by costs. These applications include high-power and high-frequency devices where silicon technology cannot meet the performance requirements such as high dynamic range or low noise figure. Figure 1.9 shows an adaptation of a graph that they published showing a prediction of where the current and future technologies might lie. GaN occupies a very large proportion of this graph and in recent years, research into this area has accelerated which can probably be attributed to the realization that the market potential will be so large. A novel GaN-based high electron mobility transistor is considered which significantly increases the drive current and thereby the power efficiency. Quantum well based transistors are very well suited, as they confine the charge carriers to the high-mobility material using heterostructure isolation.

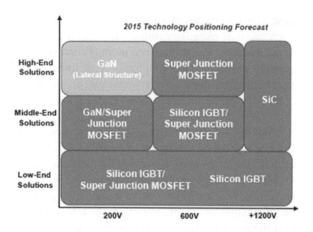

FIGURE 1.9   Forecast of potential applications of different technologies including nitride semiconductors [71].

High-mobility materials such as Ge and III-V bring promise in increasing drive current through an order of magnitude increase in intrinsic mobility. With the scaling in gate length, the impact of mobility on drain current becomes limited because of the velocity saturation. On the other hand, whenever gate length further scales down, the carrier transport becomes ballistic. This allows the velocity of carriers, also known as the "injection velocity" scale with the mobility increase. However, low effective mass for the high-mobility device can bring high tunneling current at higher supply voltage. This may degrade the performance of III-V devices at short channels after work function tuning (e.g., threshold voltage increase) to lower the leakage current (I_{off}) to compensate for the tunneling current. Another consideration for a high-mobility channel is the lower density of states. The current is proportional to the multiplication of drift velocity and carrier concentration in the channel [72]. To maximize the benefits of high-mobility channels in the drain current, it becomes much more important to reduce the contact resistance. One promising reduction is achieved by metal–insulator–semiconductor (MIS) contacts, which utilize an ultra-thin dielectric between the metal and semiconductor interface. This reduces the Fermi level pinning and therefore reduces the Schottky barrier height (SBH) [73, 74]. In this chapter, advanced devices involving III-V semiconductors are considered. A novel GaN-based high electron mobility transistor is considered which significantly increases the drive current and thereby the power efficiency.

1.4.7   CHAPTER 8 STRAIN-ENGINEERED SIGE CHANNEL TFTS FOR FLEXIBLE ELECTRONICS

A new field of nanoscience and technology based on flexible nanomembranes is rapidly developing [75], with the realization that the excellent electronic properties of strained single-crystal Si need not be confined solely to the bulk rigid substrates of conventional Si-based devices. Flexible large-area electronics are an area of increasing research interest [76] and offer a wealth of potential applications. Flexible films have the inherent advantage of being lightweight and immune to shattering, making them well suited for implementation in portable devices. Flexibility also provides new freedoms in terms of design by enabling the development of conformal or flexible displays and sensors. Examples of applications under development include wearable health monitors; flexible electronic readers [77]; contact-lens displays [78]; hemispherical imaging arrays that allow for wide-angle viewing [79]. Anticipating a limit to this continuous miniaturization (More-Moore), intense research efforts have been invested to co-integrate various functionalities (More-than-Moore). In this context, flexible and even stretchable electronics [80] offer many opportunities.

A common component to both display and imaging applications is the thin-film transistor (TFT). TFTs are used in active-matrix backplanes, which allow for highly responsive and precise control of the arrays of lighting or sensor elements that form display or imaging screens. A major challenge of developing flexible displays and sensors with amorphous silicon is electrical metastability, such as threshold voltage, which gradually shifts overtime when a gate bias is applied. The metastability leads to a degradation in the operation of TFT circuits and limits the device lifetime. Flexible devices can experience different types of deformations such as tension, compression, shear, bend, and torsion. As a general approach to deal with any combination of those deformation modes, we can apply prescribed motions at boundaries. In this chapter, a nonlinear numerical deformation stress simulation TCAD framework has been developed and applied to SiGe channel TFT design on a flexible substrate. The tensile strain or compress strain by uniaxial bending stress is demonstrated to take into account the effects on the electrical performance. We have demonstrated a simple and viable approach to realizing strained-SiGe channel RF transistors on flexible substrates. This technique has great potential in low-power and high-speed flexible-electronics applications and could be used to replace several rigid counterparts for use in mechanically bendable and non-planar conformal surfaces where rigid devices cannot be easily integrated. One can foresee as a consequence manufacturable large-area applications of such flexible high-speed TFT technology.

1.4.8 SUMMARY

In this chapter, we have presented a comprehensive overview of the monograph. The scaling challenges CMOS is facing were described in detail, including the trade-off between power dissipation and performance; the vertical and lateral scaling challenges such as gate direct tunneling and short channel effects; reliability and variability. Finally, the technology boosters, such as stress engineering and high-k/metal gates, all employed to enable continued scaling were presented. Each chapter provides the basis of understanding needed for the next chapters of the monograph. As the monograph encompasses broad areas of semiconductor technology, physics, modeling, and characterization, and the monograph being planned as a design book, sources have been referred for interested readers for detail. Simulation-based scaling study – including device design, performance characterization, and the impact of variability on nanoscale devices is considered. This monograph is intended both as a self-learning resource or professional reference and as a text for use in a graduate course. A key feature of the monograph is a modular organization. The chapters may be read in almost any order. Most of the simulation results presented throughout this monograph were performed using Silvaco and GTS Framework.

REFERENCES

[1] R. H. Dennard et al., "Design of ion-implanted MOSFET's with very small physical dimensions," *IEEE J. Solid-State Circuits*, vol. 9, no. 5, pp. 256–268, 1974, doi: 10.1109/JSSC.1974.1050511.

[2] G. E. Moore, "Cramming more components onto integrated circuits," *Electronics*, vol. 38, no. 8, p. 114, 1965, doi: 10.1109/N-SSC.2006.4785860.

[3] X. Klemenschits, S. Selberherr, and L. Filipovic, "Modeling of gate stack patterning for advanced technology nodes: A review," *Micromachines*, vol. 9, no. 12, 2018, doi: 10.3390/mi9120631.

[4] M. Vinet et al., *"FDSOI nanowires: An opportunity for hybrid circuit with field effect and single electron transistors,"* in *2013 IEEE International Electron Devices Meeting*, 2013, pp. 26.4.1–26.4.4, doi: 10.1109/IEDM.2013.6724697.

[5] J. G. Fossum et al., *"Physical insights on design and modeling of nanoscale FinFETs,"* in *IEEE IEDM Tech. Dig.*, 2003, pp. 679–682, doi: 10.1109/IEDM.2003.1269371.

[6] P. Batude et al., *"Advances, challenges and opportunities in 3D CMOS sequential integration,"* in *2011 International Electron Devices Meeting*, 2011, pp. 7.3.1–7.3.4, doi: 10.1109/IEDM.2011.6131506.

[7] IEEE, "International roadmap for devices and systems, 2020 update, more Moore." 2020.

[8] G. E. Moore, *"Progress in digital integrated electronics,"* in *IEEE IEDM Tech. Dig.*, 1975, pp. 11–13, doi: 10.1109/N-SSC.2006.4804410.

[9] G. E. Moore, *"No exponential is forever: But 'Forever' can be delayed! [semiconductor industry],"* in *2003 IEEE International Solid-State Circuits Conference, 2003. Digest of Technical Papers. ISSCC*, 2003, pp. 20–23, vol. 1, doi: 10.1109/ISSCC.2003.1234194.

[10] IEEE, "International roadmap for devices and systems," 2018.

[11] D. Bergeron, *"More than Moore,"* in *2008 IEEE Custom Integrated Circuits Conference*, 2008, pp. xxv–xxvi, doi: 10.1109/CICC.2008.4672003.

[12] A. Loke and J. Lai, *"Session 15 - IC technology - more Moore and more than Moore,"* in *2008 IEEE Custom Integrated Circuits Conference*, 2008, pp. liv–lv, doi: 10.1109/CICC.2008.4672099.

[13] A. B. Kahng, "Scaling: More than Moore's law," *IEEE Des. Test Comput.*, vol. 27, no. 3, pp. 86–87, 2010, doi: 10.1109/MDT.2010.71.

[14] N. Collaert, *"More Moore: From device scaling to 3D integration and system-technology co-optimization,"* in *2017 Silicon Nanoelectronics Workshop (SNW)*, 2017, pp. 123–124, doi: 10.23919/SNW.2017.8242328.

[15] T.-B. Chan, A. B. Kahng, and J. Li, *"Reliability-constrained die stacking order in 3DICs under manufacturing variability,"* in *International Symposium on Quality Electronic Design (ISQED)*, 2013, pp. 16–23, doi: 10.1109/ISQED.2013.6523584.

[16] S. Bansal et al., *"3-D stacked die: Now or future?"* in *Design Automation Conference*, 2010, pp. 298–299, doi: 10.1145/1837274.1837350.

[17] J. R. Hu, J. Chen, B. Liew, Y. Wang, L. Shen, and L. Cong, *"Systematic co-optimization from chip design, process technology to systems for GPU AI chip,"* in *2018 International Symposium on VLSI Technology, Systems, and Application (VLSI-TSA)*, 2018, pp. 1–2, doi: 10.1109/VLSI-DAT.2018.8373280.

[18] M. Bohr, "The evolution of scaling from the homogeneous era to the heterogeneous era," 2011, doi: 10.1109/IEDM.2011.6131469.

[19] G. Yeap, *"Technology-design-manufacturing co-optimization for advanced mobile SoCs,"* in *Proceedings of the IEEE 2014 Custom Integrated Circuits Conference*, 2014, pp. 1–8, doi: 10.1109/CICC.2014.6946000.

[20] C. K. Maiti, *Introducing Technology Computer-Aided Design (TCAD) Fundamentals, Simulations, and Applications*. CRC Press (Taylor and Francis), Pan Stanford, USA, 2017.

[21] C. K. Maiti, S. Chattopadhyay, and L. K. Bera, *Strained-Si Heterostructure Field-Effect Devices*. CRC Press (Taylor and Francis), Pan Stanford, USA, 2007.

[22] M. Heyns et al., *"Advancing CMOS beyond the Si roadmap with Ge and III/V devices,"* in *2011 International Electron Devices Meeting*, 2011, pp. 13.1.1–13.1.4, doi: 10.1109/IEDM.2011.6131543.

[23] N. Planes et al., *"28 nm FDSOI technology platform for high-speed low-voltage digital applications,"* in *Digest of Technical Papers - Symposium on VLSI Technology*, 2012, pp. 133–134, doi: 10.1109/VLSIT.2012.6242497.

[24] C. Auth et al., *"A 22nm high performance and low-power CMOS technology featuring fully-depleted tri-gate transistors, self-aligned contacts, and high density MIM capacitors,"* in *Digest of Technical Papers - Symposium on VLSI Technology*, 2012, pp. 131–132, doi: 10.1109/VLSIT.2012.6242496.

[25] J. Lacord, "Models development for power performance assessment of advanced CMOS technologies sub-20nm." PhD Thesis, Univ. of Grenoble, 2012.

[26] S. Takagi et al., "Carrier-transport-enhanced channel CMOS for improved power consumption and performance," *IEEE Trans. Electron Devices*, vol. 55, no. 1, pp. 21–39, 2008, doi: 10.1109/TED.2007.911034.

[27] D. Bae et al., *"A novel tensile Si (n) and compressive SiGe (p) dual-channel CMOS FinFET co-integration scheme for 5nm logic applications and beyond,"* in *2016 IEEE International Electron Devices Meeting (IEDM)*, 2016, pp. 28.1.1–28.1.4, doi: 10.1109/IEDM.2016.7838496.

[28] O. Weber et al., "High immunity to threshold voltage variability in undoped ultra-thin FDSOI MOSFETs and its physical understanding," 2008, doi: 10.1109/IEDM.2008.4796663.

[29] O. Weber et al., *"Gate stack solutions in gate-first FDSOI technology to meet high performance, low leakage, VT centering and reliability criteria,"* in *2016 IEEE Symposium on VLSI Technology*, 2016, pp. 1–2, doi: 10.1109/VLSIT.2016.7573434.

[30] M. Badaroglu et al., *"PPAC scaling enablement for 5nm mobile SoC technology,"* in *2017 47th European Solid-State Device Research Conference (ESSDERC)*, 2017, pp. 240–243, doi: 10.1109/ESSDERC.2017.8066636.

[31] A. Veloso et al., *"Challenges and opportunities of vertical FET devices using 3D circuit design layouts,"* in *2016 IEEE SOI-3D-Subthreshold Microelectronics Technology Unified Conference (S3S)*, Burlingame, California, 2016, pp. 1–3.

[32] T. P. Ma, *"Beyond Si: Opportunities and challenges for CMOS technology based on high-mobility channel materials,"* in *Sematech Symposium*, Taiwan, 2012.

[33] T. Skotnicki and F. Boeuf, *"How can high mobility channel materials boost or degrade performance in advanced CMOS,"* in *2010 Symposium on VLSI Technology*, 2010, pp. 153–154, doi: 10.1109/VLSIT.2010.5556208.

[34] K. J. Kuhn, A. Murthy, R. Kotlyar, and M. Kuhn, "Past, present, and future: SiGe and CMOS transistor scaling," *ECS Trans.*, vol. 33, no. 6, pp. 3–17, 2019, doi: 10.1149/1.3487530.

[35] R. Berthelon et al., *"A novel dual isolation scheme for stress and back-bias maximum efficiency in FDSOI Technology,"* in *2016 IEEE International Electron Devices Meeting (IEDM)*, 2016, pp. 17.7.1–17.7.4.

[36] R. Carter et al., *"22nm FDSOI technology for emerging mobile, Internet-of-Things, and RF applications,"* in *2016 IEEE International Electron Devices Meeting (IEDM)*, 2016, pp. 2.2.1–2.2.4, doi: 10.1109/IEDM.2016.7838029.

[37] S. C. Song et al., *"Holistic technology optimization and key enablers for 7nm mobile SoC,"* in *2015 Symposium on VLSI Circuits (VLSI Circuits)*, 2015, pp. T198–T199, doi: 10.1109/VLSIC.2015.7231373.

[38] K. Cheng et al., *"Air spacer for 10nm FinFET CMOS and beyond,"* in *2016 IEEE International Electron Devices Meeting (IEDM)*, 2016, pp. 17.1.1–17.1.4, doi: 10.1109/IEDM.2016.7838436.

[39] N. Collaert et al., *"Beyond-Si materials and devices for more Moore and more than Moore applications,"* in *2016 International Conference on IC Design and Technology (ICICDT)*, 2016, pp. 1–5, doi: 10.1109/ICICDT.2016.7542050.

[40] L. Filipovic and S. Selberherr, *"Integration of gas sensors with CMOS technology,"* in *2020 4th IEEE Electron Devices Technology Manufacturing Conference (EDTM)*, 2020, pp. 1–4, doi: 10.1109/EDTM47692.2020.9117828.

[41] T. Magis et al., "Wafer-level 3-D integration moving forward," *Semicond. Int.*, 2006.

[42] J. U. Knickerbocker et al., "Three-dimensional silicon integration," *IBM J. Res. Dev.*, vol. 52, pp. 553–569, 2008, doi: 10.1147/JRD.2008.5388564.

[43] G. G. Shahidi, "SOI technology for the GHz era," *IBM J. Res. Dev.*, vol. 46, pp. 121–132, 2002, doi: 10.1147/rd.462.0121.

[44] K. Oshima et al., "Advanced SOI MOSFETs with buried alumina and ground plane: Self-heating and short-channel effects," *Solid. State. Electron.*, vol. 48, no. 6, pp. 907–917, 2004, doi: 10.1016/j.sse.2003.12.026.

[45] F. Gamiz et al., "Electron mobility in extremely thin single-gate silicon-on-insulator inversion layers," *J. Appl. Phys.*, vol. 86, pp. 62–69, 1999, doi: 10.1063/1.371684.

[46] G. K. Celler and S. Cristoloveanu, "Frontiers of silicon-on-insulator," *J. Appl. Phys.*, vol. 93, pp. 4955–4978, 2003, doi: 10.1063/1.1558223.

[47] P. Ramm et al., "3D System-on-Chip technologies for More than Moore systems," *Microsyst. Technol.*, vol. 16, p. 1051–1055, 2010, doi: 10.1007/s00542-009-0976-1.

[48] F. Bacchini et al., *"Megatrends and EDA 2017,"* in *2007 44th ACM/IEEE Design Automation Conference*, 2007, pp. 21–22, doi: 10.1109/DAC.2007.375045.

[49] G. A. Armstrong and C. K. Maiti, *TCAD for Si, SiGe, and GaAs Integrated Circuits*. The Institution of Engineering and Technology (IET), UK, 2008.

[50] Global TCAD Solutions GmbH, "GTS framework." 2018.

[51] Silvaco International, "Silvaco suite." 2018.

[52] J. Lee et al., *"Nanowire FETs,"* in *2018 IEEE 13th Nanotechnology Materials and Devices Conference (NMDC)*, 2018, pp. 1–4, doi: 10.1109/NMDC.2018.8605884.

[53] C. Medina-Bailon et al., *"Study of the 1D scattering mechanisms' impact on the mobility in Si nanowire transistors,"* in *2018 Joint International EUROSOI Workshop and International Conference on Ultimate Integration on Silicon (EUROSOI-ULIS)*, 2018, pp. 1–4, doi: 10.1109/ULIS.2018.8354723.

[54] O. Badami et al., "Comprehensive study of cross-section dependent effective masses for silicon based gate-all-around transistors," *Appl. Sci.*, vol. 9, no. 9, 2019, doi: 10.3390/app9091895.

[55] X. Garros et al., "Modeling and direct extraction of band offset induced by stress engineering in silicon-on-insulator metal-oxide-semiconductor field effect transistors: Implications for device reliability," *J. Appl. Phys.*, vol. 105, no. 11, p. 114508, 2009, doi: 10.1063/1.3126506.

[56] S. Thompson et al., *"A 90 nm logic technology featuring 50 nm strained silicon channel transistors, 7 layers of Cu interconnects, low k ILD, and 1 um2 SRAM cell,"* in *Technical Digest - International Electron Devices Meeting*, 2002, pp. 61–64, doi: 10.1109/iedm.2002.1175779.

[57] M. Yang et al., "Hybrid-Orientation Technology (HOT): Opportunities and challenges," *IEEE Trans. Electron Dev.*, vol. 53, pp. 965–978, 2006, doi: 10.1109/TED.2006.872693.

[58] K. Shin, C. O. Chui, and T. J. King, *"Dual stress capping layer enhancement study for hybrid orientation FinFET CMOS technology,"* in *Technical Digest - International Electron Devices Meeting, IEDM*, 2005, vol. 2005, pp. 988–991, doi: 10.1109/iedm.2005.1609528.

[59] Silvaco International, VictoryStress User Manual, 2018.

[60] K. Kuhn, "Chapter 1 - CMOS and beyond CMOS: Scaling challenges," *High Mobility Materials for CMOS Applications*. Woodhead Publishing, pp. 1–44, 2018. doi:10.1016/B978-0-08-102061-6.00001-X.

[61] C. Sampedro, F. Gámiz, and A. Godoy, "On the extension of ET-FDSOI roadmap for 22 nm node and beyond," *Solid. State Electron.*, vol. 90, pp. 23–27, 2013, doi: 10.1016/j.sse.2013.02.057.

[62] C. Auth et al., *"A 10nm high performance and low-power CMOS technology featuring 3rd generation FinFET transistors, Self-Aligned Quad Patterning, contact over active gate and cobalt local interconnects,"* in *2017 IEEE International Electron Devices Meeting (IEDM)*, 2017, pp. 29.1.1–29.1.4, doi: 10.1109/IEDM.2017.8268472.

[63] T. P. Dash, S. Dey, S. Das, E. Mohapatra, J. Jena, and C. K. Maiti, "Strain-engineering in nanowire field-effect transistors at 3 nm technology node," *Phys. E Low-dimensional Syst. Nanostructures*, vol. 118, p. 113964, 2020, doi: 10.1016/j.physe.2020.113964.

[64] T. P. Dash, J. Jena, E. Mohapatra, S. Dey, S. Das, and C. K. Maiti, "Stress-induced variability studies in tri-gate FinFETs with source/drain stressor at 7 nm technology nodes," *J. Electron. Mater.*, vol. 48, no. 8, pp. 5348–5362, 2019, doi: 10.1007/s11664-019-07348-7.

[65] M. S. Bhoir et al., *"Process-induced Vt variability in nanoscale FinFETs: Does Vt extraction methods have any impact?"* in *2020 4th IEEE Electron Devices Technology Manufacturing Conference (EDTM)*, 2020, pp. 1–4, doi: 10.1109/EDTM47692.2020.9117815.

[66] T. P. Dash, S. Dey, J. Jena, S. Das, E. Mohapatra, and C. K. Maiti, *"Metal grain granularity induced variability in gate-all-around Si-nanowire transistors at 1nm technology node,"* in *2019 Devices for Integrated Circuit (DevIC)*, 2019, pp. 286–290, doi: 10.1109/DEVIC.2019.8783717.

[67] S. M. Nawaz, S. Dutta, and A. Mallik, "A comparison of random discrete dopant induced variability between Ge and Si junctionless p-FinFETs," *Appl. Phys. Lett.*, vol. 107, no. 3, 2015, doi: 10.1063/1.4927279.

[68] A. R. Brown, N. M. Idris, J. R. Watling, and A. Asenov, "Impact of metal gate granularity on threshold voltage variability: A full-scale three-dimensional statistical simulation study," *IEEE Electron Device Lett.*, vol. 31, no. 11, pp. 1199–1201, 2010, doi: 10.1109/LED.2010.2069080.

[69] H. Dadgour et al., "Statistical modeling of metal-gate work-function variability in emerging device technologies and implications for circuit design," *Proc. IEEE/ACM Int. Conf. Comput. Des.*, pp. 270–277, 2008, doi: 10.1109/ICCAD.2008.4681585.

[70] H. F. Dadgour, K. Endo, V. K. De, and K. Banerjee, "Grain-orientation induced work function variation in nanoscale metal-gate transistors - Part I: Modeling, analysis, and experimental validation," *IEEE Trans. Electron Devices*, vol. 57, no. 10, pp. 2504–2514, 2010, doi: 10.1109/TED.2010.2063191.

[71] D. J. Macfarlane, "Design and fabrication of AlGaN/GaN HEMTs with high breakdown voltages." PhD thesis, University of Glasgow, 2014.

[72] S. Wu et al., *"A 7nm CMOS platform technology featuring 4th generation FinFET transistors with a 0.027um2 high density 6-T SRAM cell for mobile SoC applications,"* in *2016 IEEE International Electron Devices Meeting (IEDM)*, 2016, pp. 2.6.1–2.6.4, doi: 10.1109/IEDM.2016.7838333.

[73] G. Eneman et al., "*Stress simulations for optimal mobility group IV p- and nMOS FinFETs for the 14 nm node and beyond*," in *2012 International Electron Devices Meeting*, 2012, pp. 6.5.1–6.5.4, doi: 10.1109/IEDM.2012.6478991.

[74] R. Xie et al., "*A 7nm FinFET technology featuring EUV patterning and dual strained high mobility channels*," in *2016 IEEE International Electron Devices Meeting (IEDM)*, 2016, pp. 2.7.1–2.7.4, doi: 10.1109/IEDM.2016.7838334.

[75] H. Jang, T. Das, W. Lee, and J.-H. Ahn, "Transparent and foldable electronics enabled by Si nanomembranes," *Silicon Nanomembranes*. John Wiley & Sons, Ltd, pp. 57–88, 2016, doi: 10.1002/9783527691005.ch3.

[76] J. Sanz-Robinson et al., "Large-area electronics: A platform for next-generation human-computer interfaces," *IEEE J. Emerg. Sel. Top. Circuits Syst.*, vol. 7, no. 1, pp. 38–49, 2017, doi: 10.1109/JETCAS.2016.2620474.

[77] M. Banks, "Flexible electronics enters the e-reader market," *Phys. World*, vol. 23, no. 02, p. 8, Feb. 2010, doi: 10.1088/2058-7058/23/02/14.

[78] J. N. Burghartz, W. Appel, C. Harendt, H. Rempp, H. Richter, and M. Zimmermann, "Ultra-thin chip technology and applications, a new paradigm in silicon technology," *Solid-State Electronics*, 2010, vol. 54, no. 9, pp. 818–829, doi: 10.1016/j.sse.2010.04.042.

[79] R. H. Reuss et al., "Macroelectronics: Perspectives on technology and applications," *Proc. IEEE*, vol. 93, no. 7, pp. 1239–1256, 2005, doi: 10.1109/JPROC.2005.851237.

[80] W. Dong, L. Yang, and G. Fortino, "Stretchable human machine interface based on smart glove embedded with PDMS-CB strain sensors," *IEEE Sens. J.*, vol. 20, no. 14, pp. 8073–8081, 2020, doi: 10.1109/JSEN.2020.2982070.

2 Simulation Environment

Technology Computer-Aided Design (TCAD) tools are widely used in industrial environments for process modeling and predictive device simulation of the electrical characteristics of microelectronics components. This is mainly because TCAD device design and process technology can not only reduce cost and development time but also secure technology choice [1]. A reliable TCAD tool is also required for supporting the physics coherence and performance analysis of highly scaled devices, for detecting operating limits, and also for investigating new device concepts. Nevertheless, the monograph is associated with the recommendations of the IRDS roadmap [2] concerning the efforts made by industry and academic partners in the modeling of future generations of transistors. To create a foundation for the simulation environment and to ensure reproducibility, the used material models and tools are introduced in the beginning.

In this chapter, currently available advanced semiconductor device/process simulation tools are introduced. The evolution of semiconductor simulation tools goes back to the late 1960s. Starting in the late 1960s, the process and device simulation were predominantly in 1D and 2D. Device simulation is dominantly 2D due to the nature of devices and became the workhorse in the design and scaling of devices. The transition from n-MOSFET to CMOS technology resulted in the necessity of tightly coupled and full 2D simulators for process and device simulations. Due to the ultra-small size of the state-of-the-art devices, the 3D effects became dominant. To achieve a better understanding of simulated and fabricated device characteristics, the 3D process/device simulation has now been introduced. Currently, there are several commercial TCAD software platforms for process/device simulation around the world. Most of them were first developed by the universities worldwide and later integrated into a suite by the TCAD vendors. Since the early 1990s, commercial vendors such as SILVACO and SYNOPSYS developed their graphical user interfaces (GUIs) around existing frameworks, which facilitated the integration of process and device simulation tools for use by non-specialists within a wider engineering environment. The concept of general-purpose process and device simulators that allows flexible simulation of different structures in different technologies became a reality.

Since the 1990s, there has been an increasing demand for simulators that comprehensively provide the capability of modeling for as many different types of semiconductor devices and processes as possible [3]. This goal can either be achieved by incorporating new features in a simulator or by establishing links between simulators with different capabilities, such as coupling of electronic design (ECAD) with TCAD. TCAD originated from bipolar technology in the late 1960s, solving the 1D and 2D process control issues. With the advent of computer modeling of semiconductors in the 1970s and the popularity of MOSFET technologies in the 1980s, tremendous efforts were put into the development of the TCAD [4]. The goal of TCAD is to offer interactive physical description and investigation of semiconductor devices by simulation and modeling to support circuit design as well as to reduce R&D time and cost. Using TCAD, conventional semiconductor device characteristics can be predicted with time and economic efficiency based on the developed model and simulation algorithm, while novel process, structure, and materials can be visualized through phenomenological and semi-empirical models to catch the physical insights and speed up the learning curve [5].

Currently available commercial simulation tools for semiconductor device and process simulations are SDEVICE and SPROCESS from Synopsys, ATHENA, and ATLAS from SILVACO. For model parameter extractions, AURORA from Synopsys, and UTMOST from SILVACO are the most commonly used tools. ATLAS is a device simulator, which allows users to create a 2D structure of a semiconductor device, including the definition of oxide and silicon regions, doping profiles to simulate the current-voltage characteristics of the device. ATLAS features include transient and AC small-signal analysis, impact ionization, gate current, and ionization integrals. There are different transport models available in device simulators, for example, Drift-Diffusion (DD) and hydrodynamic models. ATLAS is a general-purpose optimization tool for fitting analytical models to data and extract parameters for circuit simulation.

Commercial tools are being designed to address the aforementioned challenges by combining the best-in-class features with a wide range of new features arising out of new technology generations. Typical requirements for the TCAD simulation GUI are: (1) easy to use – new users should be able to pick up the basics within minutes, (2) real-time help for the command syntax, and (3) integration with other GUIs of the software suite, like the 3D setting up the tool and plotting GUI. In the semiconductor industry, this is useful because optimizing device performance often requires to adjusting process parameters. There may be hundreds of possible process step combinations, so it is useful to automate the process as much as possible.

2.1   SYNOPSYS TCAD TOOLS

The "Sentaurus TCAD" toolset, provided by Synopsys, is a comprehensive nanoscale process, device design, and simulation tools. It supports industry-leading process and device simulation with a powerful GUI-driven simulation environment for managing simulation tasks and analyzing simulation results. Synopsys Sentaurus TCAD is used for the semiconductor device and circuit design, characterization, modeling, and analysis [6]. The design and simulation flow in the Sentaurus TCAD framework is shown in Figure 2.1. The device geometry structure with a doping profile is designed by either Sentaurus Process or Sentaurus Structure Editor. The former is a 3D capable process simulator for semiconductor process technology development and optimization. Its comprehensive process models cover implantation, diffusion, annealing, etching, oxidation, epitaxial growth, etc. The latter is not only a 3D-capable device editor but also a 3D-capable process emulator. And geometric and process emulation operations can be mixed freely, adding more flexibility to the generation of 3D structures. The Sentaurus Structure Editor is suitable for the design of non-classical devices such as novel silicon transistors and non-silicon devices. The device

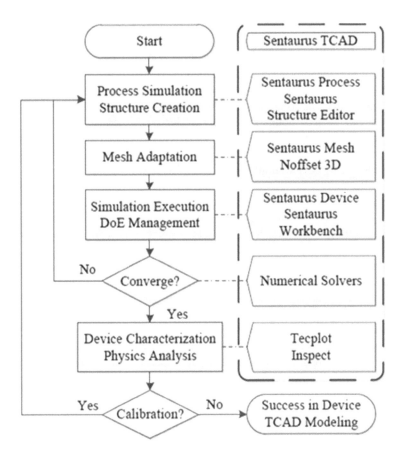

FIGURE 2.1 Sentaurus TCAD simulation flow.

created by either Sentaurus Structure Editor or Sentaurus Process is a "virtual" device whose physical properties are assigned to a finite number of discrete grids of nodes. This grid adaptation procedure is completed by Sentaurus Mesh, which is a mesh generator that incorporates two mesh generation engines: an axis-aligned mesh generation engine and a tensor-product engine that produces rectangular or hexahedral elements.

The choice of the appropriate mesh generator depends largely on the geometry of the device and the essential surfaces within the device. For planar devices such as conventional MOSFET and layer-stack HEMT, an axis-aligned mesh generator is recommended. For a device where most important surfaces are non-axis-aligned or curved, the tensor-product engine Noffset3D is adopted to produce meshes containing layers to better conform to the curved surfaces. The mesh quality is controlled by the refinement information according to user requirements. Generally, a total node count of 2,000 to 4,000 is reasonable for most 2D simulations. 3D structures require a considerably larger number of elements. It is noted that the mesh quality will have an impact on simulation accuracy, efficiency, and robustness. A most suitable mesh strategy should compromise fine mesh for high current density, high electric fields, high charge generation regions to ensure the accuracy and robustness, and coarse mesh to relatively low physics activity regions such as substrate and most source/drain regions for the improvement of simulation efficiency. After the device is passed by the mesh engine, the next steps include applications of physics models, bias condition, and numerical solution algorithm in Sentaurus Device, a comprehensive, semiconductor device simulator capable of electrical, thermal, and optical device characteristics simulation. Similar to other TCAD toolsets, Sentaurus TCAD solves fundamental equations along with its advanced model in explaining non-ideal device phenomena during device simulation.

Four different carrier transport models are available for the description of current density in various applications. The Drift-diffusion model can be selected in low-power density long channel device isothermal simulation. The Thermodynamic model accounting for the self-heating effect is suitable for a device with a low thermal exchange. The Hydrodynamic model can be applied in a small active region device and the Monte Carlo model is capable of a full-band structure solution. In this work, we employ first the carrier transport models and select the appropriate one for device simulation according to the device material, structure and operation, and interested device characteristics.

In addition to basic transport equations, TCAD modeling of semiconductor devices also need to address the band structure and the mobility properties. The band structure models are involved with bandgap, electron affinity, and effective mass (EFM), and effective density of states. The mobility models discuss the impact of phonon scattering, impurity ion scattering, carrier–carrier scattering, and surface roughness and scattering. When we simulate devices other than MOSFET, particular models for material and physical phenomena should be included in the simulation. For instance, tunneling and trapping must be recorded in the simulation of a nonvolatile memory whereas the spontaneous and piezoelectric model should be introduced in the investigation of heterostructure devices. Besides general simulation methods, phenomenological and semi-empirical models are often used to account for novel processes, material, or structure and to compromise the simulation efficiency and robustness. For example, the high-k metal gate stack can be modeled through changes to the equivalent oxide thickness with an additional dipole layer; the Ohmic contact on the nitride materials can be visualized as a Schottky contact with only electron tunneling to improve the simulation convergence rate.

The simulation convergence, as well as simulation time, also depends on the iterative algorithm settings of the Newton solver. Several factors are needed to be addressed for the trade-offs among simulation efficiency, accuracy, and robustness: the maximum number of iterations; the desired precision of the solution; the linear solver appropriate for interested device operation; and the introduction of damping methods in the expedition of the initial solution search at the cost of result accuracy. To achieve a good convergence rate of simulations, efforts need to put on the compromise among physics model, mesh quality, and solution algorithm. It should be pointed out that examination of the simulated device physics properties such as electron density, electric field, and electrostatic potential will help identify the cause of the convergence issue.

2.2   SILVACO TCAD TOOLS

SILVACO provides a comprehensive set of electronic design automation (EDA) software, which allows semiconductor companies to design both analog and mixed-signal integrated circuits. SILVACO delivers products like TCAD tools for process and device simulation. It provides analog semiconductor process, device, and design automation solutions in CMOS, bipolar transistors, SiGe, and many other compound semiconductor technologies. The main process and the device simulation tools are the ATHENA and ATLAS for the process, device simulation, and characterization, respectively.

2.2.1   DECKBUILD

Interactive Deck Development and Runtime Environment DECKBUILD is an interactive runtime and input file development environment within which all Silvaco's TCAD and several other SIMUCAD products can run. A brief overview of Silvaco TCAD software is shown in Figure 2.2. DECKBUILD has numerous simulator-specific and general debugger-style tools. This includes powerful extract statements, GUI-based process file input, and line-by-line runtime execution, and intuitive input file syntactical error messages. DECKBUILD contains an extensive library of hundreds of pre-run example decks that cover many technologies and materials and also allow the user to rapidly become highly productive. Key features of DECKBUILD are:

- Provide an interactive runtime and input file development environment for running several core simulators
- Input deck creation and editing

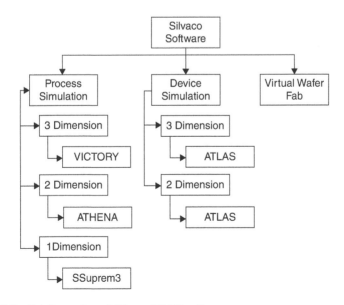

FIGURE 2.2   Brief overview of Silvaco TCAD software.

- View simulator output and controls
- Set popups that provide full language and run-time support for each simulator
- The automatic interface among simulators
- Many input file creation and debug assist features, such as run, kill, pause, stop at, and restart
- Extracted quantities can be used as targets in DECKBUILD's internal optimizer, allowing automatic cyclical optimization of any parameter.

2.2.2 Silvaco ATLAS

ATLAS enables one to simulate the electrical, optical, and thermal behavior of semiconductor devices. ATLAS provides a physics-based, easy-to-use, modular, and extensible platform to analyze DC, AC, and time-domain responses for semiconductor devices in 2D and 3D. Accurately characterizes physics-based devices in 2D or 3D for electrical, optical, and thermal performance without costly split-lot experiments. Solves yield and process variation problems for the optimal combination of speed, power, density, breakdown, leakage, luminosity, or reliability. Figure 2.3 shows the ATLAS Framework [7]. ATLAS can conduct simulation-based on Quantum theories, magnetic effect, lattice heating, and many more. Collectively, these features allow the researcher to conduct a wide variety of simulations related to modern semiconductor technologies. ATLAS device simulation program for 2D and 3D utilizes the same program, but depending on the degrees of dimensions, different smaller programs are packaged into ATLAS as presented in Figure 2.3.

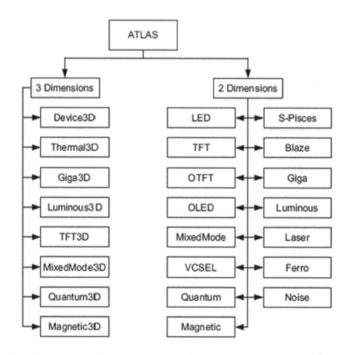

FIGURE 2.3   Silvaco ATLAS Framework Architecture.

2.2.3 SILVACO ATHENA

The software used in the project is originally from SILVACO's International. It is divided into three parts, which are the ATHENA software, used for fabrication and simulation of the device, and ATLAS software, used for investigation of the characterization of the device, and MASKVIEWS – to design a complicated MOSFET by drawing the layout. TCAD tools are chosen to aid the project since the cost to fabricate the real design is very expensive. Thus, virtual fabrication is needed, especially when the design does not meet the right specification and one has to do it all over again. If a real fabrication is done, the cost to fabricate two to three times would be so much expensive. Thus, virtual fabrication is the most suitable method to investigate the characterization of the device, without any costs. Any improvement in the design can be made without hassle with the aid of the software.

ATHENA has evolved from a world-renowned Stanford University simulator SUPREM-IV, with many new capabilities developed in collaboration with dozens of academic and industrial partners [7]. SSuprem3 is used for 1D projects while ATHENA employs SSuprem4 and a collection of other programs as presented in Figure 2.4. Fully integrated ATHENA process simulation software includes several comprehensive visualization packages and an extensive database of examples. One can choose from the largest selection of silicon, III-V, II-VI, IV-IV, or polymer/organic technologies including CMOS, bipolar, high-voltage power device, VCSEL, TFT, optoelectronic, LASER, LED, CCD, sensor, fuse, NVM, ferroelectric, SOI, Fin-FET, HEMT, and HBT technologies. ATHENA is used for process simulation and the user is required to input several parameters following ATHENA's requirements, as presented in Figure 2.4.

Silvaco provides Virtual Wafer Fab (VWF) tool involving process simulation, device simulation, process characterization, etc. A brief overview of Silvaco's TCAD software is presented in Figure 2.2. The VWF software allows split parameters to be

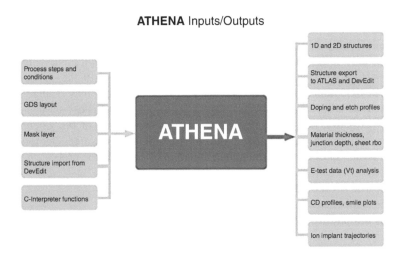

FIGURE 2.4 Silvaco ATHENA inputs and outputs.

defined for any of the Silvaco processes, device, and circuit simulators, and all simulations are carried out in parallel either on a cluster of workstations or on a single machine [8]. This allows either multiple computers to parallel compute a project or to use a single computer attached to multiple processors as the process simulation software depends on the degree of dimension simulated. In addition to this software, Silvaco provides small software that can be used as peripheral tools for conducting simulations. These include TonyPlot, TonyPlot3D, DeckBuild, MaskViews, and DevEdit.

2.2.4 VICTORYPROCESS

VictoryProcess is a fully 3D process simulator which allows user to perform a wide range of device structures of the desired shape. Most of the supported operations correspond to real processes such as etching or deposition, CMP, epitaxy, and others so that the process engineer can establish a direct link between the technological processes and an input statement for VictoryProcess. VictoryProcess can simulate complicated full physics-based etching and deposition processes. Advanced process simulation may be performed very quickly and accurately using either in the "geometrical" mode or very simple physical models. VictoryProcess can simulate complicated full physics-based etching and deposition processes that take into account; reactor characteristics (particle flux), shading effects and/or, and secondary effects like redeposition of the etched material. The etching and deposition process characteristics include:

- Rounded corners
- Tapered sidewall
- Non-conformal epitaxial growth
- Selective geometrical etch
- Selective CMP

VictoryProcess is capable of simulating complex processes within a feature scale for 3D structures (see Figure 2.5). The technological parameters from the reactor can be included using transport characteristics. Different crystal planes of silicon are known to have different oxidation rates, for example, the silicon plane with Miller indices <111> is oxidized approximately 1.7 times faster than the <100> plane. During an oxidation process, the geometry of a 3D structure may change significantly and the silicon/oxide interface may pass through various crystal planes with different oxidation rates. Therefore, the orientation and type of the silicon wafer affect the resulting geometry of an oxidized structure on this wafer. VictoryProcess offers several modes for the simulation/emulation of an oxidation process, namely,

- the analytical oxidation mode
- the empirical oxidation mode
- the full physical oxidation mode

Since the FEM provides an approximate solution to the stress analysis problem, the result will have some dependence on the mesh used. First of all, the solution to

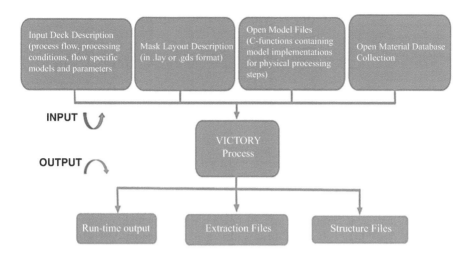

FIGURE 2.5   The input and output arguments of VictoryProcess.

the problem is calculated in the nodes defined by the mesh. Fewer nodes mean having fewer elements where the solution is calculated. Therefore, it is necessary to include as many elements as possible to improve the accuracy. It is also necessary to have the elements as regularly as possible to enhance the accuracy. If the element is too skinny, we will have regions with many closed solutions and regions with solutions spread apart. This makes the resulting system of equations badly conditioned, and in turn, the solution becomes less accurate especially when we use an iterative solver. VictoryProcess provides statistics on mesh quality (average and maximum element aspect ratio) for each material region and gives a warning message when the average element aspect ratio exceeds 1:100 in any region. It is recommended that the mesh space in one direction is not too small compared to other directions. VictoryProcess supports multiple meshing algorithms for 3D stress simulation.

2.2.5   VICTORYDEVICE

VictoryDevice is a general-purpose 3D device simulator. A tetrahedral meshing engine is used for fast and accurate simulation of complex 3D geometries. VictoryDevice performs DC, AC, and transient analysis for silicon, binary, ternary, and quaternary material-based devices (see Figure 2.6).

VictoryDevice features include:

- Tetrahedral mesh for accurate 3D geometry representation
- Voronoi discretization for conformal Delaunay meshes
- Advanced physical models with user-customizable material database
- Stress-dependent mobility and bandgap models
- Physical models using the C-Interpreter or dynamically linked libraries
- DC, AC, and transient analysis

FIGURE 2.6   Input and output arguments of VictoryDevice.

- Drift-Diffusion and energy balance transport equations
- Self-consistent simulation of self-heating effects including heat generation, heat flow, lattice heating, heat sinks, and temperature-dependent material parameters
- Advanced multithreaded numerical solver library
- Atlas-compatible

2.2.6 VICTORYSTRESS

Accurate simulation of mechanical stress/strain distributions generated during device fabrication is an important part of technology and device design. In many cases, stress effects should be taken into account to predict better manufacturability and to increase the reliability of semiconductor devices. In recent years, stress simulation has become a critical issue in TCAD due to advances in stress engineering. Stress engineering could be described as a collection of device optimization methods based on the deliberate introduction of stresses into the device cell structure. These methods include Dual Stress Liners (DSLs), strain SiGe channels, SiGe (SiC) S/D pockets, the hybrid orientation of n-MOSFET and p-MOSFET devices, and an etch stop layer. All these methods are used to improve device performance by altering carrier mobility in specific areas of the device structure. VictoryStress is a generic 3D stress simulator (see Figure 2.7) that allows accurate prediction of stresses generated during semiconductor fabrication as well as assists users in all

FIGURE 2.7   VictoryStress information flow.

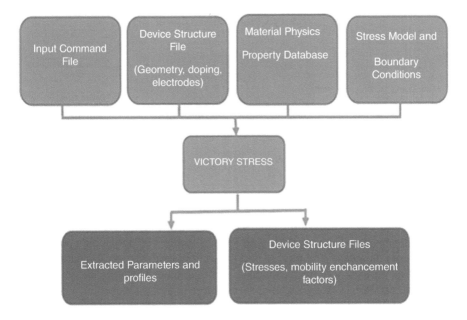

FIGURE 2.8.   Program flow in VictoryStress [9].

aspects of stress engineering. The three main modules in VictoryStress are input file generation, 3D VictoryStress simulation, and output generation. Program flow in VictoryStress is shown in Figure 2.8.

VictoryStress provides a comprehensive set of models and capabilities covering various aspects of stress simulation and stress engineering:

- Stress analysis of arbitrary 3D device structure.
- Import of device structures from VictoryProcess, VictoryCell, and ATLAS3D.
- Export of 3D stress distributions to VictoryDevice and ATLAS3D.
- Models for various sources of strain and stress
- Thermal mismatch between material layers.
- Local lattice mismatch due to doping.
- Initial intrinsic stress in specified regions.
- Hydrostatic stress from capping layers.
- Stress/strain generated in the previous processing step (e.g., oxidation).
- Stress simulation for various crystalline (e.g., Si, SiGe, GaAs) and isotropic (e.g., silicon nitride and oxide) materials.
- Generic 3D anisotropic stress simulation accounts for wafer orientation and arbitrary wafer flat rotation.
- Estimation of mobility enhancement factors (p- and n-type) by the use of piezoresistivity model devices.
- Can be used with VWF for Design of Experiments to analyze stress dependence on the process and geometrical parameters of semiconductor

Victory Cell Inputs/Outputs

FIGURE 2.9　VictoryCell input/output flow.

2.2.7　VICTORYCELL

VictoryCell is a 3D process simulator (see Figure 2.9) designed to be a very fast way of creating devices. The choice of mesh formats that can be output are optimized to be as device simulator friendly as possible, minimizing device simulation times. VictoryCell an ideal tool for the simulation of numerous technologies, such as SiC IGBT and MOSFETs.

Key features include:

- Layout-driven device creation, using either GDSII or Silvaco layout formats
- Fast 3D process modeling of etching, deposit, implant, diffusion, and photolithography
- Easy to learn and user-friendly SUPREM-like syntax
- Interfaces to both Silvaco's 3D device simulators, Device-3D and VictoryDevice

GDSII Layout Driven Process Simulation has the following advantages:

- Layout-driven simulation allows the creation of high aspect ratio 3D structures
- Simulates realistic etch and deposition steps very efficiently with unstructured tetrahedral mesh
- 3D Monte Carlo ion implantation takes full account of crystal orientation and ion stopping including the effects of damage and amorphization

2.2.8　MASKVIEWS

Another interesting part of Silvaco TCAD tools is MASKVIEWS. MASKVIEWS is an IC layout editor. It is designed to interface the IC layout or any complicated structured device with Silvaco's process simulator. MASKVIEWS can be used extensively to draw and edit any complicated device and IC layout, store and then load the complete layout, and import or export the layout details by using the industry-standard GDSII and CIF layout format. Any part of the layout can be simulated – the most interesting process, and without the hassle of structuring the codes in ATHENA; one can obtain an accurate design based on the layout that has been accurately designed. MASKVIEWS also provides features to allow layout experimentation such as misalignment, polygon oversizing or under-sizing, global rescales, and region definition – depending on combinations of present mask elements.

2.3 VSP SIMULATION SOFTWARE

VSP is a general-purpose device simulator for arbitrary nanostructures, operating on the Schrödinger–Poisson equation system. VSP uses a finite volume discretization scheme, thus avoiding the weak formulation fundamental to finite elements and relying instead on a formulation based on the conservation of fluxes in each of the finite volumes. As a quantum-electronic simulator, VSP includes quantum mechanical solvers for closed-boundary as well as open-boundary problems. The closed boundary model setup allows us to calculate a self-consistent carrier concentration in FinFET or nanowire crosssections including quantization effects due to electrical and geometrical confinement. The equations systems are solved on tetrahedral and triangular meshes for arbitrary 2D and 3D geometries. The Schrödinger solver supports EFM or KdotP Hamiltonians including arbitrary strain distribution conditions tensor. The model accounts for arbitrary substrate types and channel orientation. The effects of quantum confinement are simulated using the self-consistent coupled Schrödinger–Poisson model; however, this model cannot solve transport problems by itself. It is therefore used in combination with either the Drift-Diffusion modespace approach or the mode space non-equilibrium Green's function approach [10]. Heterostructured semiconductor devices can be treated within the closed boundary model for quick estimation of resonant energy levels. The open boundary model allows the evaluation of current-voltage characteristics. Using the experimental 2D non-equilibrium Green's functions solver, nano-MOSFETs in the ballistic operation regime can be investigated. The key features include:

Arbitrary 1D/2D/3D geometries by using unstructured meshes

Substrate and channel orientation

Multivalley single-band EFM Hamiltonian

KdotP Hamiltonian including strain effect

Calculate the sub-band ladder (Eigen energies) and wave functions (including overlaps)

Physical modeling for conductivity and mobility

Based on the Boltzmann equation (Kubo–Greenwood formula)

Includes crystal orientation and strain effects

Anisotropic band-structure

Surface roughness scattering (SRS)

Non-polar phonon scattering (ADP, ODP, inter-valley)

Polar-optical phonon scattering (POP)

Ionized impurity (Coulomb) scattering (IIS)

Alloy disorder scattering (ADS)

The simulation software used for nanoscale device modeling is based on the commercial tools discussed above. The VSP tool is a 3D self-consistent, quantum simulator based on the EFM approximation. The calculation involves a self-consistent solution of a 3D Poisson equation and a 3D Schrödinger equation with open boundary conditions at the source and drain contacts. Using the finite element method (FEM), the 3D Poisson equation is solved initially to obtain the electrostatic potential throughout the device. At the same time, the 3D Schrödinger equation is solved by a (coupled/uncoupled) mode space approach, which provides both computational efficiency and high accuracy as compared with direct real space calculations [11, 12].

2.4   MINIMOS-NT

MINIMOS-NT is the most important discrete device simulator which provides the user with the ability to simulate a broad range of fully-featured 2D and 3D devices. Minimos-NT is a general-purpose semiconductor device simulator providing steady-state, transient, and small-signal analysis of arbitrary 2D and 3D device geometries. In the mixed-mode device and circuit simulation, numerically simulated devices can be placed in circuits built from compact device models and passive elements. These simulators incorporate advanced physical models and robust numeric methods for the simulation of the most common types of semiconductor devices. Device simulation is based on solving carrier transport equations in arbitrary semiconductor structures. The device under investigation is partitioned into segments and for each segment, a material is assigned. The equations that have to be solved are assembled for each segment and the boundaries in between them, depending on the material, the active transport equations, and the activated physical models [13]. A partial differential equation system describes the device behavior using physics-based transport equations and the corresponding model parameters. For the solution of this system, a discretization in space is required, which is accomplished by defining a geometry mesh. The equations are assembled for each grid point. Such an equation system cannot be solved directly. Therefore, iterative solution methods like the Newton iteration system have to be applied. The use of complex physical models, which often add highly non-linear relations between the variables, also increases the complexity of the equation system leading to an increased number of iteration steps and therefore longer convergence times. Some setups can also lead to failed conversions. Also, the boundary conditions, for example, the given contact potentials, influence the convergence. The iterative solution process needs an initial value (called initial guess) to solve the given problem. If no initial value is available, a first initial guess is calculated using thermal equilibrium in the semiconductor. Alternatively, a previous simulation result can be used as an initial guess.

2.4.1   FINITE ELEMENT SIMULATION

Solving a physical problem usually requires one to take into account several parameters such as system geometry, material properties, and boundary conditions. When the system has a complex geometry, it is often difficult to solve the differential equations that govern it analytically. In this situation, it becomes easier to deal with the problem digitally. The FEM is based on this idea to a discretization of the problem. The basic principle of simulation is therefore to divide the geometry into a more or less important of elements and to solve in each of the nodes the physical laws. This method allows us to determine the distribution of certain variables such as displacement in the case of strain analysis, temperature, or heat flow in the case of thermal analysis for example.

The FEM simulation procedure usually consists of the following steps:

The creation of geometry. It is formed from straight lines, surfaces, or curves. It can be defined as a code, although it may be practical to have software available when the geometry is complex. The internal boundaries of geometry are generally determined by the properties of materials or media.

The mesh of the structure. The size of the mesh (or the number of elements, the two being connected), is an important parameter for the quality of the model. Solutions that minimize the energy of the system are searched at each node of the network. So, with more elements considered, the model is more accurate, but longer will be the calculation time. In practice, it is often necessary to determine the number of elements (especially in the case of 3D models) constituting a compromise between model accuracy and required computing time. The density of elements can, however, be cleverly adjusted according to the position in the model. For the cases we are interested in, the number of elements is increased in areas where the deformation is important.

The specification of the properties of the materials or the medium. For a strain analysis, it will be necessary to specify, for example, Young's moduli or the coefficients of elasticity of the materials.

For thermal analysis, we will define the thermal conductivities. Simulation software sometimes includes databases on common materials. The specification of the properties simply by assigning the parameters to a specific area of the geometry.

Definition of boundary conditions and initial conditions. The application of these conditions is a function of the problem considered. They can have a lot of influence on the result simulation.

2.5 TECHNOLOGY CAD SIMULATION

An ideal TCAD device simulation tool should be able to precisely predict changes in process and device geometry providing reliable results for ultimate technology nodes. For all these reasons, advanced device modeling strategies are required to account for many aspects, such as stress engineering, quantum effects, reliable mobility models, channel materials, and complex architectures. In the simulation-based research of aggressively scaled CMOS transistors, it is mandatory to combine advanced transport simulators and quantum confinement effects with atomistic simulations that accurately reproduce the electronic structure at the nanometer scale. Typical device geometry and dimension-dependent physical model selection are shown in Figure 2.10. In this context, more sophisticated stand-alone solvers are

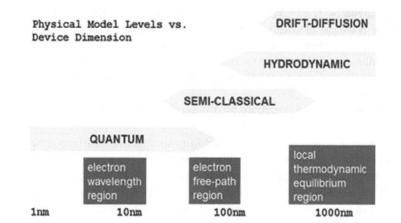

FIGURE 2.10   Physical model levels vs. device dimension.

available, such as Kubo–Greenwood [14], Multi-Subband Monte Carlo (MSMC) [15], Spherical Harmonics Expansion (SHE) [16], or NEGF solvers [17], offering an attractive alternative to DD/HD-based tools. Besides, these tools can also be used to calibrate DD/HD solvers, when users need to keep using them. Nonetheless, accurate solutions such as MSMC, SHE, or NEGF are very time-consuming. Long simulation times become a concern in the industrial development environment, where many simulation runs of complex device architectures are typically needed. For this reason, it is highly desired not only as a reliable but also as a fast simulation tool to support technology development and predict device performance.

In this section, we shall discuss various models used in device simulation tools. We shall examine the consistency of traditional TCAD tools for modeling devices at the scaling limit. Model selection hierarchy in advanced device simulation is shown in Figure 2.11. Mostly all simulations are performed starting with DD and extending to DG along with its analysis of validity and limitations. Density-Gradient (DG) model with either DD or HD transport model. Comparisons are made using simulation results obtained for the same device structure with the channel length, thickness-dependent mobility, the remote Coulomb scattering, and the nonlinear piezoelectric strain-dependent models. The details on these models and the calibration procedures on advanced solvers can be found in reference [18].

The finesse of physics and all the mathematical tools, included in the modeling will depend on the desired precision as well as the desired calculation time. Thus, modeling of material, of a transistor area, of an entire transistor, or even the modeling of a circuit do not require the same digital tools and the same physical knowledge. Indeed, the physical phenomena to be taken into account are not the same depending on the scale studied.

Device simulation tool requirements at the nanoscale:

Ultimate nanoscale device simulation capability will contribute to the prediction of CMOS limits.

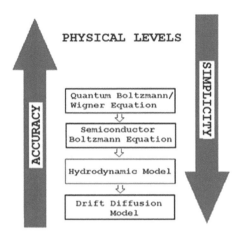

FIGURE 2.11   Model selection hierarchy in device simulation.

Device modeling is used to study the scaling limits in devices for technology optimization. The ability to correctly reproduce device performance and predict the possible limitations of the future device.

Numerical methods are required for improvements to support TCAD tools in taking into account the increasing complexity of physical phenomena.

Device Modeling is used to include the effects of stress on the band structures (displacement of bands, the density of states, EFMs).

Materials modeling used in the simulation of devices can be considered as models of materials since they are based on the electronic structure of semiconductors.

Ultimate nanoscale device simulation capability needs to include strain engineering. The efficiency, precision, and robustness of calculations are key issues.

The current complexity of component physics requires the use of Monte Carlo methods, which stochastically solve the Boltzmann equation.

The fully Depleted SOI, multigate devices require partially ballistic transport models. Description of mobility for arbitrary channel directions becomes essential. A full description of stress effects becomes necessary.

Ultimate nanoscale device simulation requires a Quantum-based simulator, precision, and efficiency in calculation time, the inclusion of band structures, and the spectrum of phonons. Simulations should apply beyond standard planar CMOS devices.

The current complexity of the physics of the components requires the use of simulators, based on the Schrödinger equation, which takes into account quantum effects.

Ultimate nanoscale device simulation needs to have predictive physical models on performance enhanced devices under stress.

In device modeling, the impact of strain on mobility is of crucial importance. They induce an anisotropic piezoresistivity, caused not only by the change in EFMs but also by the variation of the relaxation times, as well as by the dependence of the saturation rate on the strain.

2.6 DEVICE SIMULATION

Modeling is a powerful tool for engineering design and analysis which usually makes a description or analogy of a phenomenon to visualize something not directly observed. The definition of modeling may vary depending on the application, but the basic concept of modeling is the process of solving physical problems by appropriate simplification of reality. Modeling usually consists of constructing a mathematical model for corresponding physical problems with appropriate assumptions and the model may take the form of differential or algebraic equations. In many engineering cases, these mathematical models cannot be solved analytically, often requiring a numerical solution. The process of modeling generally involves the development of an appropriate physical model and it typically requires careful calibration and validation against pre-existing data and/or analytical results. The term "device modeling," for the microelectronics field, refers to a collection of physical models and methodologies describing carrier transport and other physical effects in semiconductor devices [19]. The advanced simulation of transport properties requires the description of the band structures of the component materials of the modeled device. The modeling approach, both in the calculation of band structures and in that of transport properties, consists of fitting the simple models (fast simulation) to the most predictive models (complex and therefore expensive in terms of calculation time). In the context of band structure calculations, ab initio simulations like semi-empirical EPM, k.p models allow in a few seconds to evaluate the entire electronic structure.

Over the years, device engineers have improved our collective knowledge of carrier transport and semiconductor physics. The transport models range from the so-called DD approach [20], which is extensively used in industry due to its simplicity and efficiency, to complex and computationally demanding such as semi-classical or quantum transport models. For instance, the Non-Equilibrium Green's function (NEGF) formalism [21–24] and the Monte Carlo (MC) method [25–27] are common approaches used to solve respectively, the Schrödinger and the Boltzmann transport equations with few approximations are required for an accurate description of the band structure of the semiconductor and also of the scattering mechanisms limiting the mobility.

The transport models range from the so-called Drift-Diffusion approach [20], which is extensively used in industry due to its simplicity and efficiency, to complex and computationally demanding such as semi-classical or quantum transport models. For instance, the Non-Equilibrium Green's function formalism [23, 28–30] and the Monte Carlo method [25–27] are common approaches used to solve respectively the Schrödinger and the Boltzmann transport equations with few approximations or hypotheses. For advanced transport solvers such as NEGF, MC, and Kubo–Greenwood (KG) [14, 31–33], it is required an accurate description of the band structure of the semiconductor and also of the scattering mechanisms limiting the mobility.

2.6.1 DRIFT-DIFFUSION SIMULATION

The simulation of electronic devices generally involves a self-consistent solution of the electrostatic potential and carrier distribution inside the device. Earlier treatment of electrons and holes as semi-classical particles with an EFM was good enough to predict semiconductor device behavior and the Drift-Diffusion equation was adequate to describe carrier transport. Today, as we stand at 5 nm technology nodes and devices have shrunk to nanoscale dimensions, which has required a re-examination of our approach to device modeling. In particular, the properties of materials can be altered using strain, heterojunctions and computational grading are possible, and quantum effects start to show up in the nanometer regime. As a result, the conventional Drift-Diffusion and Boltzmann equations do not capture the increasingly important role of quantum mechanics in modeling transistors in 10 nm regime.

The Boltzmann equation is particularly complex to solve in its entirety. Besides, the modeling of electronic devices does not generally require a description of detailed microscopic mechanisms of transport. Models have been developed taking into account approximations on some terms of the Boltzmann equation. These models allow flexibility of use as well as a significant reduction in computing time. We briefly recall, in this paragraph, the Drift-Diffusion models which can be deduced from the Boltzmann equation by the method of moments. Mobility is essential data in the calculation of the Drift-Diffusion model. DD model computes via analytical expressions based on semi-empirical considerations. The adjustment coefficients are calibrated using experiments or simulations based on advanced transport models (Monte Carlo, Kubo–Greenwood). In Drift-Diffusion simulations, we used the Masetti model [34], which models the reduction in mobility as a function of the increase in the concentration of carriers. At high fields, the charge carrier drift speed is no longer proportional to the electric field. It saturates at a finite value of speed. In Drift-Diffusion simulations, we used the Canali extended model.

The inversion channel in a MOSFET is strongly influenced by the gate potential. However, the strong potential gradients confine the charges near the oxide of the gate-by-gate potential. Under these polarization conditions, the wave nature of the carriers of charge dictates their probability of being present in the inversion layer. The main consequence is the distance of the barycenter of the inversion charges from the Si/oxide interface, leaving a space with almost no voltage at the Si/oxide interface. In this section, we introduce the fundamental concepts of the transport of charge from the expansion of the Boltzmann equation. Subsequently, we detail the numerical methods that are used in the calculation of transport properties: the Drift-Diffusion, Monte Carlo, and Kubo–Greenwood methods.

2.6.2 BOLTZMANN TRANSPORT MODEL

The Boltzmann equation plays a role in the theoretical understanding of transport phenomena, that is to say, the response of a system maintained in external conditions of imbalance (difference in potential, temperature, concentration, or speed, imposed between two points). Its development is particularly well detailed in references

[25, 35, 36]. We briefly recall, in the following, the models of Drift-Diffusion which can be deduced from the Boltzmann equation by the method of moments:

$$\overrightarrow{J_n}(r,t) = |q| D_n \vec{\nabla} n(r,t) + |q| n(r,t) \mu_n \vec{E} \tag{2.1}$$

$$\overrightarrow{J_p}(r,t) = -|q| D_p \vec{\nabla} p(r,t) + |q| p(r,t) \mu_p \vec{E} \tag{2.2}$$

$$D_n = \frac{kT}{|q|} \mu_n, D_p = \frac{kT}{|q|} \mu_p \tag{2.3}$$

The indices n and p in Equations (2.1–2.3) are respectively associated with electrons and holes. D represents the diffusion coefficient. μ corresponds to mobility.

In the expressions of the Drift-Diffusion model, the continuity Equations:

$$\frac{\partial n}{\partial t} n(r,t) - \frac{1}{|q|} \vec{\nabla} \cdot \overrightarrow{J_n} - s_n(r,t) = 0 \tag{2.4}$$

$$\frac{\partial p}{\partial t} p(r,t) - \frac{1}{|q|} \vec{\nabla} \cdot \overrightarrow{J_p} - s_p(r,t) = 0 \tag{2.5}$$

In Equations (2.4, 2.5), the indices n and p are respectively associated with electrons and holes. $n(r,t)$ and $p(r,t)$ represent the density of electrons and holes. $s(r, t)$ is related to generation rates and recombination of electron-hole pairs. Effective mobility is linked with drift velocity, which corresponds to the average speed of the charge carriers. However, in the context of ultra-short channel transistors in a high-field regime, Drift-Diffusion simulations must have currents consistent with those resulting from advanced transport models.

2.6.3 ENERGY BALANCE MODEL

Like the Drift-Diffusion model, the energy balance transport model is derived by taking moments of the Boltzmann transport equation. Besides balance equations for the carrier density and momentum, the energy balance model includes balance equations for the carrier energy and energy flux. The expression for current density can be expressed as

$$\overrightarrow{J_n} = qn\mu_n \overrightarrow{E_n} + qD_n \vec{\nabla} n + qnD_n^T \vec{\nabla} T_n \tag{2.6}$$

$$\overrightarrow{J_p} = qp\mu_p \overrightarrow{E_p} - qD_p \vec{\nabla} p - qpD_p^T \vec{\nabla} T_p \tag{2.7}$$

The energy balance (EB) model is used to account for the effects of the ballistic carrier transport. It significantly increases drain currents and is the cause of the decrease in the output impedance of the I_d-V_d curves. It also uses the Bohm Quantum Potential (BQP) model to take into account the quantum confinement of carriers in 3D which decreases the drain currents.

2.6.4 QUANTUM MECHANICAL MODEL

In the classical Drift-Diffusion transport model quantum effect is not taken into account. But as the device scales down to the nanometer range, the quantum effect comes into the picture in determining device characteristics. For this reason, a more sophisticated analysis of the device physics is needed, such as the non-equilibrium Green's function approach, to model devices to a ballistic level (<10 nm) [37]. The NEGF transport model is by far the most rigorous approach among existing quantum transport models [10, 38]. So this effect needs to be taken into account during the simulation. According to classical theory, the position of the particle has been determined with certainty but the position of the particle is expressed in terms of probability in quantum mechanics. The position of the electron is described by the probability density function. When density gradient is needed, one can also make use of the Poisson solution with the Schrödinger equation (self-consistent) for the correct calibration of the model. The behavior of an electron inside the crystal is then described by the wave function. For scaled-down devices, the quantum effects need to be considered in simulation [39]. Quantum mechanics is associated with atomic and subatomic systems. When the device size is in the nanometer range, quantum effects will play a vital role in determining device characteristics. In quantum mechanics, the wave function ψ of the particle which changes with time is given by:

$$\psi = R \exp\left(i \frac{S}{\hbar} \right) \tag{2.8}$$

where R is a probability density per unit volume, is the modulus of Ψ and S has the dimension of action (energy × time) and $\hbar$ is the reduced Planck constant.

2.6.5 BOHM QUANTUM POTENTIAL MODEL

The BQP model is used to determine the quantum mechanical effects. The BQP model introduces a position-dependent quantum potential, $\varnothing_{BQ}$ which is added to the Potential Energy of a given carrier type. The great advantage of the BQP model consists of the possibility to correctly model quantum effects without the explicit solution of the Schrödinger equation. The quantum potential in BQP Model is given by:

$$E_{BQ,n} = q\phi_{BQ,n} = \frac{-\hbar^2 \gamma_n \nabla\left(M^{-1}\nabla\left(n^{\alpha_n} \right) \right)}{2 \cdot n^\alpha} \tag{2.9}$$

$$E_{BQ,p} = q\phi_{BQ,p} = \frac{-\hbar^2 \gamma_p \nabla\left(M^{-1}\nabla\left(p^{\alpha_p} \right) \right)}{2 \cdot p^{\alpha_p}} \tag{2.10}$$

where γ and α are two adjustable parameters, M^{-1} is the inverse EFM tensor and n is the electron (or hole) density. E_{BQ} is the energy equivalent to BQP. The modified current density can be expressed as:

$$\vec{J}_n = qn\mu_n \vec{\nabla}\left(\phi_n + \vec{\nabla} E_{BQ,n} \right) \tag{2.11}$$

and

$$\overrightarrow{J_p} = qp\mu_p\vec{\nabla}\left(\phi_p - \vec{\nabla}E_{BQ,p}\right) \tag{2.12}$$

Equations (2.6, 2.7) are modified when the BQP is used and are given by:

$$\overrightarrow{J_n} = qn\mu_n\overrightarrow{E_n} + qD_n\vec{\nabla}n + qnD_n^T\vec{\nabla}T_n + qn\mu_n\vec{\nabla}E_{BQ,n}, \tag{2.13}$$

and

$$\overrightarrow{J_p} = qp\mu_p\overrightarrow{E_p} - qD_p\vec{\nabla}p - qpD_p^T\vec{\nabla}T_p - qn\mu_p\vec{\nabla}E_{BQ,p} \tag{2.14}$$

where $E_{BQ,n} = q\phi_{BQ,n}$ is the BQP for electrons, and
$E_{BQ,p} = q\phi_{BQ,p}$ is the BQP for holes. These models and their modifications discussed in this section have been implanted in device simulation and used for the analysis of electrical performance.

2.6.6 DENSITY-GRADIENT MODEL

Theories of carrier transport can be divided according to whether they are microscopic or macroscopic in character [40]. Microscopic theories deal with individual carriers, such as wave functions and density matrices. The macroscopic theories are devoted to the carrier populations. For the latter, semiconductor devices that are small enough are directly impacted by quantum mechanics effects, such as the phenomena of quantum confinement. Quantum confinement is basically due to the impact on the atomic structure as a result of the direct effect of nanoscale lengths on the energy band structure [41]. For nanostructures, quantum effects become relevant due to the ratio between the size of the device (thin film) or potential well in the case of electrical confinement and the mean free path of the carrier.

Contrary to classical charge distribution, the quantum confinement is described by envelop functions associated with a well-defined number of discrete energy levels and these envelop functions will have their maximum density pushed-away from the Si-SiO₂ interface. The low-density zone close to this interface is the so-called "dark space." For quantum confinement in the channel thickness direction, electrons are no longer represented as a 3D electron gas (3DEG), but as a 2D electron gas (2DEG) in which the transport is modeled in the 2D of the channel plane. A simplified TCAD model accounting for quantization effects is given, for instance, by the modified local-density approximation (MLDA) which is a model that calculates the confined carrier distributions occurring near semiconductor–insulator interfaces [42]. It can be applied to both inversion and accumulation regimes, and simultaneously to electrons and holes.

This model is derived from the method of moments applied to the Wigner equation [43]. The Wigner function is defined by the Fourier transform of the product of wave functions in two points in space. The expression of the density of charge carriers in the Density-Gradient model differs from the classical expression only by the presence of potential energy. The effect quantum can thus appear as the term of an additional force that is composed of higher-order derivatives of classical potential [44]. In the Density-Gradient model, the quantum correction potentials γ_n and γ_p for electrons and holes, respectively, have been added to the current relation of the DD model have been therefore extended as:

$$J_n = q \cdot \mu_n \cdot n \left[grad \left(\frac{E_C}{q} - \psi - \gamma_n \right) + \frac{k_B \cdot T_L}{q} \cdot \frac{N_{C,0}}{p} \cdot grad \left(\frac{n}{N_{C,0}} \right) \right] \qquad (2.15)$$

$$J_p = q \cdot \mu_p \cdot p \left[grad \left(\frac{E_V}{q} - \psi - \gamma_p \right) - \frac{k_B \cdot T_L}{q} \cdot \frac{N_{V,0}}{p} \cdot grad \left(\frac{p}{N_{V,0}} \right) \right] \qquad (2.16)$$

J_n and J_p represent the electron and hole current densities, respectively; μ_n and μ_p represent electron and hole mobilities, respectively; and n and p represent electron and hole concentrations, respectively. These current relations account for position-dependent band edge energies, E_C and E_V, and position-dependent EFMs, which are included in the effective density of states, $N_{C,0}$ and $N_{V,0}$. Index 0 indicates that $N_{C,0}$ and $N_{V,0}$ are evaluated at some room temperature, k_B is Boltzmann Constant, and T_L is local lattice temperature. The simplified first-order approximation of the quantum-correction potential derived from Wigner's equation [45] is given as:

$$\gamma_n = \frac{\hbar}{12 \cdot \lambda_n \cdot m_0} \cdot div\,grad \, \frac{\psi + \gamma_n - \dfrac{E_C}{q}}{k_B \cdot T_L} \qquad (2.17)$$

$$\gamma_p = \frac{\hbar}{12 \cdot \lambda_p \cdot m_0} \cdot div\,grad \, \frac{\psi + \gamma_p - \dfrac{E_V}{q}}{k_B \cdot T_L} \qquad (2.18)$$

where $\hbar$ is reduced Planck constant, λ_n and λ_p represent the wavenumber.

2.6.7 THE MONTE CARLO METHOD

The Monte Carlo method consists of statistically solving the Boltzmann equation in simulating the behavior of charge carriers in the device, without any macroscopic approximation. The solutions resulting from the Monte Carlo method are often a reference when configuring the other simulators since they remain very close to the phenomena of fundamental physics [25]. Transport, under strongly out-of-equilibrium conditions, requires the exact resolution of Boltzmann's master equation. These conditions are particularly present in nanodevices. Specific simulation tools are needed to model these out-of-equilibrium transport effects [46]. It may be noted that the band structure of the material constituting the channel is crucial data that must be provided to Monte Carlo simulators. The operation of a Monte Carlo simulator requires the integration of bands, as well as interaction rates calculated from the band structure [47]. Those data are involved in the transport of particles of the Monte

Carlo method. The Monte Carlo method is particularly suitable for solving the equation of Boltzmann. Monte Carlo simulations are used as a reference in the calibration of the methods of moments of the Boltzmann equation, more efficient in computation time (Drift-Diffusion, Hydrodynamics) [26]. Interactions of particles between particles and their environment (phonons, impurities, etc.) are calculated by the collision integral. After each interaction, a final wave vector is determined by calculating the collision integral. These interactions materialize in Monte Carlo simulations by interruptions in the trajectories of particles in real and reciprocal space. Solving the Poisson equation coupled with the distribution of particles in the system and statistics is collected at regular intervals depending on the type of simulation Monte Carlo considered [48]. Every interval, the value of the current in the considered system (drain current for devices) is averaged over all the iterations taken into account during the statistics. The convergence of Monte Carlo simulations is achieved when the difference in the mean value of the current between two Poisson iterations is less than one bar satisfactory error (user-defined).

2.6.8 Kubo–Greenwood Transport Model

The Kubo–Greenwood transport formula is based on Boltzmann's master equation, of which it is a linearization, in the context of a 3D gas. This model is suitable for both 3DEGs and 2DEGs. Several approximations of the Boltzmann transport equation lead to the linearized formula for Kubo–Greenwood. The numerical approach consists of the Kubo–Greenwood formula and the self-consistent solution of the Schrödinger and Poisson equations for cylindrical gated nanowires. Phonons and surface roughness scatterings are treated following the literature. The impact of trapped charges in the oxide, called remote Coulomb scattering, is modeled following the Kubo–Greenwood approach and accounting for the screening effect. The study of electron mobility in intrinsic silicon nanowires using the Kubo–Greenwood approach coupled to a self-consistent Schrödinger–Poisson solver allows important possibilities of investigation as to fundamental mechanisms of electronic transport in the channel. At the nanoscale, the transport regime is completely modified due to multi-subband transport. The electronic transport is investigated numerically using the Kubo–Greenwood approach to observe separately the influence of temperature and carrier concentration on electronic transport properties in the channel.

2.6.9 Mode Space Approach

The decoupled mode space approximation consists of decoupling the 2D Schrödinger equation in two 1D equations: the first for the confinement direction to determine the modes of the electronic wave functions and the second along the channel to determine the transport. This composition can also be regarded as a pseudo-2D solution to the charge density. The total charge density is computed from the wave functions given by the closed boundary 1D sliced solutions (also named "modes" indexed by increasing energies) which are combined with the charge density given by the open boundary NEGF solution along the transport direction. More details about NEGF and mode space approximation can be found in reference [11, 49].

Generally, when the channel length is larger than 0.1 μm, macroscopic variables can be applied such as electron density, velocity, energy, and electron temperature. Among semi-classical transport models appropriate for this realm, the Drift-Diffusion model, the hydrodynamic model, and the six-moment equation can be used for modeling devices with long channel lengths ($L > 0.1$ μm). For short channel lengths, a quantum mechanical description is needed and the Density Gradient and effective potential approach need to be used. When the channel length is less than 0.1 μm, the Boltzmann transport equation must be solved by the Monte Carlo method or by a direct method. At short channel lengths (<0.1 μm) the NEGF, Wigner, and Pauli master equations can be used to describe quantum effects. The conventional transport theories based on BTE focus on scattering-dominated transport; however, for nanoscale devices, these may operate in a quasi-ballistic transport regime where scattering effects become less important and carrier transport approaches purely ballistic transport. As we enter the regime of ultra-scaled channel lengths in the range of 10 nm or lower, it becomes imperative to use the NEGF transport model.

2.7   MECHANICAL STRESS MODELING

The electronic structure of materials is modified by the application of stress. Thus, the resistivity of the material depends on its state of deformation commonly known as piezoresistivity. In the framework of Si and Ge, this phenomenon was demonstrated and formalized by Smith in 1954 [50]. Piezoresistivity is an anisotropic property: the increase or decrease of resistivity depends on the orientation of the stress [51, 52]. Although there is a consensus on the value of the piezoresistance coefficients in solid materials, their values require calibration. Indeed, the values piezoresistance coefficients will depend on the nature of the transport in the architecture of the device. The calibration of these coefficients is carried out on experimental measurements, for well-determined stress values, or even on the results of advanced simulations transportation. The piezoresistance model is commonly used in Drift-Diffusion simulators for its ease of use and efficiency. We shall present the model of piezoresistance in Chapter 4. Notions on the relations between stress and strain tensors, as well as the notation adapted to their use in the piezoresistance model, shall be presented in Chapter 4.

2.7.1   MODELING APPROACHES

The definition of modeling may vary depending on the application, but the basic concept of modeling is the process of solving physical problems by appropriate simplification of reality. Modeling usually consists of constructing a mathematical model for corresponding physical problems with appropriate assumptions and the model may take the form of differential or algebraic equations. In many engineering cases, these mathematical models cannot be solved analytically, often requiring a numerical solution. The process of modeling generally involves the development of an appropriate physical model and it typically requires careful calibration and validation against pre-existing data and/or analytical results. The term "device modeling," for the microelectronics field, refers to a collection of physical models and methodologies describing carrier transport and other physical effects in semiconductor devices [1].

The electrostatic solution of the system is determined using the Poisson equation, which is essentially based on Gauss's law [53]. The Poisson equation establishes a relationship between the electrostatic potential and the electric charge density for a given material. The carrier transport models can be written in the form of the continuity equation, which describes the charge conservation. The carrier transport models differ in the expressions used to compute the carrier's current Density. Among the simplest transport models, Drift-Diffusion has been widely used in the industry essentially due to the simplicity of its use and will supposedly continue to be applied in the future, provided that the parameters are needed in the simulation are accurately calibrated [54]. Traditional TCAD tools make available transport models such as Drift-Diffusion, which treats carrier transport as diffusion and drift processes. For the drift component, the carrier movement is due to an applied electric field, while for the diffusion component, the carrier displacement is due to spatial charge variation. In the DD approach, the electron gas is assumed to be in thermal equilibrium with the lattice temperature. Nevertheless, nanoscale devices are usually under a strong electric field in which carriers gain energy from the field and therefore their temperature is raised and rather non-uniform along with the device [55]. Compared with DD, the HD model has extra driving forces accounting for energy distribution on the transport of carriers. Besides the contribution due to spatial variation of electrostatic potential and contribution due to the gradient of carrier concentration, the current density has also a contribution due to the carrier temperature gradient and the spatial variation of the EFM in heterostructure devices. The HD model is capable of handling non-uniform temperatures and changes on EFMs along with the device.

Moreover, as the device size approaches the nanometer range, carrier transport becomes quasi-ballistic, and non-local effects such as velocity overshoot occur [56]. In an attempt to capture these phenomena, more advanced transport models have been proposed, such as hydrodynamic (HD), which can also be found in TCAD solvers. Both the DD and HD models can be viewed as approximations of the Boltzmann transport equation, which represents a rigorous approach to model carrier transport in semiconductors. Different models taking into account higher moments of the BTE such as SHE can be found in the literature [57–59]. Furthermore, rather more elaborated models must be applied to account for extra features such as stress, non-parabolic bands, and geometric quantization in thin-layer structures [60], and for all these models, appropriated calibrations are needed.

The calculation of the electron concentration in the presence of confinement effect requires solving, for instance, the Poisson and Schrödinger equation self-consistently [61]. However, when DG is required, one can also make use of the self-consistent Poisson–Schrödinger solution for proper calibration of the model [62]. However, in a realistic short-channel device, the driving electric field is no longer constant along the current path and the velocity saturation concept becomes somehow irrelevant due to, for example, velocity overshoot phenomena [63]. The saturation regime of MOSFET devices is a real concern for standard TCAD simulations. In practice, the best "local" driving force to be used in a device simulation remains unclear.

2.7.2 Schrödinger–Poisson Simulation

In the nanodevices transport simulation framework, approaches that incorporate important quantum effects into semi-classical models have become very popular due to their lower computational cost in comparison to the purely quantum transport simulation techniques. The carrier charge distribution is resolved by the nonlinear Poisson equation, although in a classical manner. The charge description for nanodevices is controlled by a quantum mechanical size quantization and for this reason, the Schrödinger equation is needed to accurately compute the quantum charge distribution. In this section, we present the evaluation of the Schrödinger equation for 1D, 2D, and 3D Gas. The Schrödinger equation in its general form, including the time dependency, is given as [64]:

$$-\frac{\hbar^2}{2}\nabla_r\left(\frac{1}{m_{eff}}\nabla_r\Psi(r,t)\right)+V(r)\Psi(r,t)=i\hbar\frac{\partial\Psi(r,t)}{\partial t} \tag{2.19}$$

where V is the time-independent potential energy of an electron at point r. The wave functions are expressed as:

$$\Psi(r,t)=\phi(r)\xi(t) \tag{2.20}$$

Thus, Equation (2.19) can also be expressed as:

$$-\frac{\hbar^2}{2\phi(r)}\nabla_r\left(\frac{1}{m_{eff}}\nabla_r\phi(r)\right)+V(r)=\frac{i\hbar}{\xi(t)}\frac{\partial\xi(t)}{\partial t} \tag{2.21}$$

The right-hand side term of Equation (2.21) depends only on the spatial coordinates, and the right-hand side term depends only on the time. These terms are equal and therefore constants. Being ε this constant, the expression in Equation (2.19) is equivalent to the following set of equations:

$$-\frac{\hbar^2}{2}\nabla_r\left(\frac{1}{m_{eff}}\nabla_r\phi(r)\right)+V(r)\phi(r)=\varepsilon\phi(r) \tag{2.22}$$

$$i\hbar\frac{\partial\xi(t)}{\partial t}=\varepsilon\xi(t) \tag{2.23}$$

where ε is the total energy of the system. The analogy between Equation (2.22) and the classical energy conservation may be noted:

$$p^2/2m_{eff}+E_p=E. \tag{2.24}$$

From the solution of Equation (2.23), the general wave function is given by:

$$\Psi(r,t)=\phi(r)e^{-i\varepsilon t/\hbar}$$

where $\phi(r)$ are the eigenfunctions of Equation (2.22).

2.7.3 THE DENSITY OF STATES 3D

In the case of a bulk semiconductor, as the crystal has a periodicity, there is no confinement. For the EFM approximation, the Schrödinger equation solution is given by plane waves as:

$$\phi(r) = \phi_0 e^{-i(k-k_0).r} \tag{2.25}$$

Considering the EFM matrix as diagonal, the Schrödinger equation is given by:

$$\frac{\hbar^2}{2}\left(\frac{(k_x - k_{0x})^2}{m_x} + \frac{(k_y - k_{0y})^2}{m_y} + \frac{(k_z - k_{0z})^2}{m_z}\right) = E - V \tag{2.26}$$

We define the ellipsoid in vector space containing the energy states that are inferior or equal to E. Its volume is given by:

$$v_E = \frac{4}{3}\pi\left(\frac{\sqrt{2m_x(E-V)}}{\hbar}\frac{\sqrt{2m_y(E-V)}}{\hbar}\frac{\sqrt{2m_z(E-V)}}{\hbar}\right) = \frac{8\sqrt{2}}{3\hbar^3}\pi\sqrt{m_x m_y m_z}(E-V)^{\frac{3}{2}} \tag{2.27}$$

The periodicity conditions are:

$$(k_x - k_{0x})d_x = 2\pi n_x, (k_y - k_{0y})d_y = 2\pi n_y, (k_z - k_{0z})d_z = 2\pi n_z \tag{2.28}$$

where n_x, n_y, and n_z are integers.

Each energy state occupies a volume of:

$$v_k = \frac{8\pi^3}{d_x d_y d_z} \tag{2.29}$$

The number of energy states that are inferior to E for the unit of volume is then (spin degeneracy included) given by:

$$N^{3D}(E) = \frac{2\sqrt{2}}{3\pi^2\hbar^3}\sqrt{m_x m_y m_z}(E-V)^{\frac{3}{2}} \tag{2.30}$$

The density of states for the unit of volume and unit of energy is given as:

$$D^{3D}(E) = \frac{\sqrt{2}}{\pi^2\hbar^3}(m^{3D})^{\frac{3}{2}}(E-V)^{\frac{1}{2}} \tag{2.31}$$

where the density of state mass m^{3D} is evaluated as follows:

$$m^{3D} = (m_x m_y m_z)^{\frac{1}{3}} \tag{2.32}$$

2.7.4 CARRIER DENSITY 3D

For each ellipsoid, the occupied density of states for the unit of volume is given by:

$$n^{3D} = \int_{E=V}^{\infty} D^{3D}(E) f\left(\sqrt{E - E_F}\right) dE \tag{2.33}$$

with the Fermi function $f(\varepsilon)$

$$f(\varepsilon) = \frac{1}{1 + \exp^{(\varepsilon)/K_B T}} \tag{2.34}$$

Thus, Equation (2.33) can be expressed as:

$$n^{3D} = \frac{\sqrt{2}\left(m^{3D}\right)^{3/2}}{\pi^2 \hbar^3} \int_{E=V}^{\infty} \frac{(E-V)^{1/2}}{1 + \exp^{[(E-V)-(E_F-V)]/K_B T}} dE \tag{2.35}$$

The corresponding Fermi integral, $\mathcal{F}_{1/2}$ is:

$$\mathcal{F}_{1/2}(\eta_F) = \frac{2}{\sqrt{\pi}} \int_{\eta=0}^{\infty} \frac{\eta^{1/2}}{1 + \exp^{(\eta - \eta_F)}} d\eta \tag{2.36}$$

Therefore, the carrier density n^{3D} is given by:

$$n^{3D} = 2\left[\frac{m^{3D} K_B T}{2\pi \hbar^2}\right]^{3/2} \mathcal{F}_{1/2}\left(\frac{E_F - V}{K_B T}\right) \tag{2.37}$$

2.7.5 THE DENSITY OF STATES 2D

When the semiconductor crystal has a periodicity for the intervals dx and dy, the density of states is presented for the case of a semiconductor with confinement in the direction z. For the EFM approximation, the Schrödinger equation's solution is given as:

$$\phi(x,y,z) = \phi_{\perp}(z) e^{i\left[(k_x - k_{0x})x + (k_y - k_{0y})y\right]} \tag{2.38}$$

Thus, the Schrödinger equation can be separated into two expressions as:

$$-\frac{\hbar^2}{2m_z} \frac{\partial^2 \phi_{\perp}}{\partial z^2}(z) + V(z)\phi_{\perp}(z) = E_{\perp}(z)\phi_{\perp}(z) \tag{2.39}$$

$$\frac{\hbar^2}{2}\left(\frac{(k_x - k_{0x})^2}{m_z} + \frac{(k_y - k_{0y})^2}{m_y}\right) = E - E_{\perp} \tag{2.40}$$

For each subband (for the component of energy $E_\perp$ in the direction of confinement), the wave vector space ellipse occupied by the energy states that are inferior or equal to E has an area:

$$S_E = \pi \frac{\sqrt{2m_x \left(E - E_\perp\right)}}{\hbar} \frac{\sqrt{2m_y \left(E - E_\perp\right)}}{\hbar} = \frac{2\pi}{\hbar^2} \sqrt{m_x m_y} \left(E - E_\perp\right) \tag{2.41}$$

and the periodicity conditions are:

$$\left(k_x - k_{0x}\right) d_x = 2\pi n_x, \left(k_y - k_{0y}\right) d_y = 2\pi n_y \tag{2.42}$$

where n_x and n_y are integers. Each energy state occupies a surface of:

$$s_k = \frac{4\pi^2}{d_x d_y} \tag{2.43}$$

The number of energy states that are inferior to E for the unit of area is, therefore (spin degeneracy included) given by:

$$N^{2D}\left(E\right) = \frac{1}{\pi \hbar^2} \sqrt{m_x m_y} \left(E - E_\perp\right) \tag{2.44}$$

The density of states for the unit of area and unit of energy is given as:

$$D^{2D}\left(E\right) = \frac{1}{\pi \hbar^2} \left(m^{2D}\right) \tag{2.45}$$

where the density of state mass m^{2D} is:

$$m^{2D} = \left(m_x m_y\right)^{1/2} \tag{2.46}$$

2.7.6 CARRIER DENSITY 2D

For each ellipsoid, the occupied density of states for the unit area is given by:

$$n^{2D} = \int_{E=E_\perp}^{\infty} D^{2D}\left(E\right) f\left(E - E_\perp\right) dE \tag{2.47}$$

Thus, with the Fermi function given by Equation (2.34), expression (2.47) can be stated as:

$$n^{2D} = \frac{m^{2D}}{\pi \hbar^2} \int_{E=E_\perp}^{\infty} \frac{1}{1 + \exp^{\left[(E-E_\perp)-(E_F-E_\perp)\right]/K_B T}} dE \tag{2.48}$$

The corresponding Fermi integral $\mathcal{F}_0$ is:

$$\mathcal{F}_0\left(\eta_F\right) = \int_{\eta=0}^{\infty} \frac{1}{1 + \exp^{(\eta-\eta_F)}} d\eta = \ln\left[1 + \exp^{(\eta_F)}\right] \tag{2.49}$$

Therefore, the carrier density n^{2D} for each ellipse is expressed as:

$$n^{2D} = 2 \left[\frac{m^{2D} K_B T}{2\pi \hbar^2} \right] \mathcal{F}_0 \left(\frac{E_F - E_\perp}{K_B T} \right)$$ (2.50)

If the semiconductor crystal has a periodicity for the interval d_x, the density of states is presented for the case of a semiconductor with confinement in the directions y and z.

2.7.7 THE DENSITY OF STATES 1D

For the EFM approximation, the Schrödinger equation solution is given by:

$$\phi(x,y,z) = \phi_\perp(y,z) e^{i\left[(k_x - k_{0x})x\right]}$$ (2.51)

Thus, the Schrödinger equation can be separated into two expressions:

$$-\frac{\hbar^2}{2m_y} \frac{\partial^2 \phi_\perp}{\partial y^2}(y,z) - \frac{\hbar^2}{2m_z} \frac{\partial^2 \phi_\perp}{\partial z^2}(y,z) + V(y,z)\phi_\perp(y,z) = E_\perp(y,z)\phi_\perp(y,z)$$ (2.52)

$$-\frac{\hbar^2}{2} \frac{(k_x - k_{0x})^2}{m_x} = E - E_\perp$$ (2.53)

For each subband (for the component of energy $E_\perp$ in the section of confinement), the wave vector space segment occupied by the energy states that are inferior or equal to E has a length:

$$L_E = \frac{2\sqrt{2m_x(E - E_\perp)}}{\hbar} = \frac{2\sqrt{2}}{\hbar} \sqrt{m_x} (E - E_\perp)^{\frac{1}{2}}$$ (2.54)

And the periodicity condition is given by:

$$(k_x - k_{0x})d_x = 2\pi n_x$$ (2.55)

where n_x is an integer. Each energy state occupies a length of:

$$l_k = \frac{2\pi}{d_x}$$ (2.56)

The number of energy states that are inferior to E for the unit of area is, therefore (spin degeneracy included):

$$N^{1D}(E) = \frac{2\sqrt{2}}{\pi \hbar} \sqrt{m_x} (E - E_\perp)^{\frac{1}{2}}$$ (2.57)

The density of states for the unit of area and unit of energy is given as:

$$D^{1D}(E) = \frac{\sqrt{2}}{\pi \hbar} \sqrt{m^{1D}} (E - E_\perp)^{-\frac{1}{2}}$$ (2.58)

where the density of state mass m^{1D}, is:

$$m^{1D} = m_x \qquad (2.59)$$

2.7.8 CARRIER DENSITY 1D

For each ellipsoid, the occupied density of states for the unit of length is:

$$n^{1D} = \int_{E=E_\perp}^{\infty} D^{1D}(E) f(E - E_F) dE \qquad (2.60)$$

Thus, with the Fermi function given by Equation (2.34), Equation (2.60) can be written as:

$$n^{1D} = \frac{\sqrt{2}\sqrt{m^{1D}}}{\pi\hbar} \int_{E=E_\perp}^{\infty} \frac{(E - E_\perp)^{-\frac{1}{2}}}{1 + \exp^{[(E-E_\perp)-(E_F-E_\perp)]/K_BT}} dE \qquad (2.61)$$

The corresponding Fermi integral $\mathcal{F}_{-\frac{1}{2}}$ is :

$$\mathcal{F}_{-\frac{1}{2}}(\eta_F) = \frac{1}{\sqrt{\pi}} \int_{\eta=0}^{\infty} \frac{\eta^{-\frac{1}{2}}}{1 + \exp^{(\eta - \eta_F)}} d\eta \qquad (2.62)$$

Therefore, the carrier density n^{1D}, for each ellipsoid is given as:

$$n^{1D} = 2 \left[\frac{m^{1D} K_B T}{2\pi\hbar^2} \right]^{\frac{1}{2}} \mathcal{F}_{-\frac{1}{2}} \left(\frac{E_F - E_\perp}{K_B T} \right) \qquad (2.63)$$

The implementation of the so-called 1.5D Poisson–Schrödinger consists of coupling a 2D nonlinear Poisson equation to a set of 1D sliced Schrödinger equations transversal to the channel direction. This composition of the slices can be regarded as a pseudo-2D solution for the Schrödinger equation.

For the 1D Schrödinger slice, the charge density is composed as:

$$n^{3D}(r) = \sum_{v,\sigma,i} \left| \psi^{v,\sigma,i}(r) \right|^2 \frac{m^{2D} K_B T}{2\pi\hbar^2} \mathcal{F}_0 \left(\frac{E_F - E_\perp^{v,\sigma,i}}{K_B T} \right) \qquad (2.64)$$

where σ is the spin degeneracy, v is the valley degeneracy, and i is the subband index.

2.7.9 SUMMARY

TCAD has been indicated by the IRDS as one of the enabling methodologies that can support the advance of technology progress at the remarkable pace of Moore's law by reducing development cycle times and costs involved in the semiconductor industry. TCAD approaches yielding accurate physical insight and useful predictive results for real-world semiconductor applications are described using multidimensional (2D and 3D) simulations. In this chapter, a brief background on past and present simulation development activities has been given. An overview of currently available commercial TCAD tools, viz., GTS Framework, Silvaco, and Synopsys, has been presented. Nevertheless, the reader should be aware that several other comprehensive simulation programs are in existence today. Developer groups in academia and industry have developed their own set of simulation programs. Indeed, several University Groups have developed programs and released them for public use. Each of these programs has certain advantages over the other regarding specific physical models, the precise mathematical equations solved, the numerical algorithms used, and the pre-and post-processing handling of data.

The weakness of traditional TCAD tools shown in this chapter requires a prompt interaction from industrial device simulation actors. TCAD carrier transport modeling is mostly based on empirical mobility models. In this context, more sophisticated solvers are available, such as Kubo–Greenwood, MSMC, SHE, or NEGF solvers, offering an attractive alternative to DD/HD-based tools. Besides, these tools can also be used to calibrate DD/HD solvers, when users need to keep using them.

Nonetheless, accurate solutions such as MSMC, SHE, or NEGF are very time-consuming. Long simulation times become a concern in the industrial environment, where many simulation runs of complex device architectures are typically needed. For this reason, it is highly desired not only as a reliable but also as a fast simulation tool to support technology development and predict device performance. As a response to all these issues, simulation tools are needed to be developed in the framework of self-consistent Poisson–Schrödinger for the electrostatics, Quantum Drift-Diffusion, and Mode Space approaches for solving the carrier transport models, Kubo–Greenwood model to account for scattering, etc. By comparing experimental data with traditional TCAD simulation results, it is important to assess not only the physical model capabilities but also its limitations which permit us to correctly respond for further modeling improvements.

REFERENCES

[1] G. A. Armstrong and C. K. Maiti, *TCAD for Si, SiGe, and GaAs Integrated Circuits*. The Institution of Engineering and Technology (IET), UK, 2008.

[2] IEEE, "International roadmap for devices and systems," 2018.

[3] C. K. Maiti, *Computer Aided Design of Micro- and Nanoelectronic Devices*. World Scientific, Singapore, 2016.

[4] R. W. Dutton and Z. Yu, *Technology CAD — Computer simulation of IC processes and devices*, Springer US, 1993, doi: 10.1007/978-1-4615-3208-8.

[5] C. K. Maiti, *Introducing Technology Computer-Aided Design (TCAD) Fundamentals, Simulations, and Applications*. CRC Press (Taylor and Francis), Pan Stanford, USA, 2017.

[6] Y.-C. Wu and Y.-R. Jhan, *3D TCAD Simulation for CMOS Nanoeletronic Devices*, Springer, 2018.

[7] Silvaco International, Santa Clara, Atlas User Manual, 2019.

[8] Silvaco International, VWF Interactive Tools User's Manual, 2018.

[9] Silvaco International, VictoryStress User Manual, 2018, 2018.

[10] O. Sikder and P. J. Schubert, *"First principle and NEGF based study of silicon nano-wire and nano-sheet for next generation FETs: Performance, interface effects and life-time,"* in *2020 IEEE 20th International Conference on Nanotechnology (IEEE-NANO)*, 2020, pp. 140–145, doi: 10.1109/NANO47656.2020.9183675.

[11] D. A. Lemus, J. Charles, and T. Kubis, "Mode-space-compatible inelastic scattering in atomistic nonequilibrium Green's function implementations," *J. Comput. Electron.*, 2020, doi: 10.1007/s10825-020-01549-8.

[12] A. Abramo, *"A general purpose 2D Schrodinger solver with open/closed boundary conditions for quantum device analysis,"* in *SISPAD '97. 1997 International Conference on Simulation of Semiconductor Processes and Devices. Technical Digest*, 1997, pp. 105–108, doi: 10.1109/SISPAD.1997.621347.

[13] MINIMOS-NT Users Guide, 2019.

[14] J. Dura, F. Triozon, S. Barraud, D. Munteanu, S. Martinie, and J. L. Autran, "Kubo-Greenwood approach for the calculation of mobility in gate-all-around nanowire metal-oxide-semiconductor field-effect transistors including screened remote Coulomb scattering-Comparison with experiment," *J. Appl. Phys.*, 2012, vol. 111, no. 10, p. 103710, doi: 10.1063/1.4719081.

[15] S. Guarnay, F. Triozon, S. Martinie, Y. Niquet, and A. Bournel, *"Monte Carlo study of effective mobility in short channel FDSOI MOSFETs,"* in *2014 International Conference on Simulation of Semiconductor Processes and Devices (SISPAD)*, 2014, pp. 105–108, doi: 10.1109/SISPAD.2014.6931574.

[16] G. Matz, S. Hong, and C. Jungemann, *"Spherical harmonics expansion of the conduction band for deterministic simulation of SiGe HBTs with full band effects,"* in *2010 International Conference on Simulation of Semiconductor Processes and Devices*, 2010, pp. 167–170, doi: 10.1109/SISPAD.2010.5604540.

[17] S. Berrada et al., "Nano-electronic Simulation Software (NESS): A flexible nano-device simulation platform," *J. Comput. Electron.*, vol. 19, no. 3, pp. 1031–1046, 2020, doi: 10.1007/s10825-020-01519-0.

[18] O. Nier, "Development of TCAD modeling for low field electronics transport and strain engineering in advanced Fully Depleted Silicon On Insulator (FDSOI) CMOS transistors," PhD Thesis, Grenoble Alpes University, 2015.

[19] O. Nier et al., "Multi-scale strategy for high-k/metal-gate UTBB-FDSOI devices modeling with emphasis on back bias impact on mobility," *J. Comput. Electron.*, vol. 12, no. 4, pp. 675–684, 2013, doi: 10.1007/s10825-013-0532-1.

[20] W. Van Roosbroeck "Theory of the flow of electrons and holes in germanium and other semiconductors," *Bell Syst. Tech. J.*, vol. 29, no. 4, pp. 560–607, 1950, doi: 10.1002/j.1538-7305.1950.tb03653.x.

[21] G. Albareda, J. Suñé, and X. Oriols, "Monte Carlo simulations of nanometric devices beyond the 'mean-field' approximation," *J. Comput. Electron.*, vol. 7, no. 3, pp. 197–200, 2008, doi: 10.1007/s10825-008-0185-7.

[22] S. Poli, M. G. Pala, T. Poiroux, S. Deleonibus, and G. Baccarani, "Size dependence of surface-roughness-limited mobility in silicon-nanowire FETs," *IEEE Trans. Electron. Devices*, vol. 55, no. 11, 2008, doi: 10.1109/TED.2008.2005164.

[23] K. Rogdakis, S. Poli, E. Bano, K. Zekentes, and M. G. Pala, "Phonon-and surface-roughness-limited mobility of gate-all-around 3C-SiC and Sinanowire FETs," *Nanotechnology*, vol. 20, no. 29, p. 295202, 2009, doi: 10.1088/0957-4484/20/29/295202.

[24] A. Martinez et al., "A self-consistent full 3-D real-space NEGF simulator for studying nonperturbative effects in nano-MOSFETs," *IEEE Trans. Electron Devices*, vol. 54, no. 9, pp. 2213–2222, 2007, doi: 10.1109/TED.2007.902867.

[25] C. Jacoboni, L. Reggiani, "The Monte Carlo method for the solution of charge transport in semiconductors with applications to covalent materials," *Rev. Mod. Phys.*, vol. 55, no. 3, pp. 645–705, 1983, doi: 10.1103/RevModPhys.55.645.

[26] M. V. Fischetti and S. E. Laux, "Monte Carlo analysis of electron transport in small semiconductor devices including band-structure and space-charge effects," *Phys. Rev. B*, vol. 38, no. 14, pp. 9721–9745, 1988, doi: 10.1103/PhysRevB.38.9721.

[27] D. Querlioz et al., "On the ability of the particle Monte Carlo technique to include quantum effects in nano-MOSFET simulation," *IEEE Trans. Electron Devices*, vol. 54, no. 9, pp. 2232–2242, 2007, doi: 10.1109/TED.2007.902713.

[28] Y. Niquet, C. Delerue, G. Allan, and M. Lannoo, "Method for tight-binding parametrization: Application to silicon nanostructures," *Phys. Rev. B - Condens. Matter Mater. Phys.*, vol. 62, no. 8, pp. 5109–5116, 2000, doi: 10.1103/PhysRevB.62.5109.

[29] S. Poli and M. G. Pala, "Channel-length dependence of low-field mobility in silicon-nanowire FETs," *IEEE Electron Device Lett.*, vol. 30, no. 11, pp. 1212–1214, 2009, doi: 10.1109/LED.2009.2031418.

[30] A. Martinez, A. R. Brown, A. Asenov, and N. Seoane, "*A comparison between a fully-3D real-space versus coupled mode-space NEGF in the study of variability in gate-all-around Si nanowire MOSFET*," in *2009 International Conference on Simulation of Semiconductor Processes and Devices*, 2009, pp. 1–4, doi: 10.1109/SISPAD.2009.5290218.

[31] O. Nier et al., "*Limits and improvements of TCAD piezoresistive models in FDSOI transistors*," in *2013 14th International Conference on Ultimate Integration on Silicon (ULIS)*, 2013, pp. 61–64, doi: 10.1109/ULIS.2013.6523491.

[32] D. Esseni and F. Driussi, "A quantitative error analysis of the mobility extraction according to the Matthiessen rule in advanced MOS transistors," *IEEE Trans. Electron Devices*, vol. 58, no. 8, pp. 2415–2422, 2011, doi: 10.1109/TED.2011.2151863.

[33] R. Kubo, "Statistical'mechanical theory of irreversible processes. I. General theory and simple applications to magnetic and conduction problems," *J. Phys. Soc. Japan*, vol. 12, no. 6, pp. 570–586, 1957, doi: 10.1143/JPSJ.12.570.

[34] G. Masetti et al., "Modeling of carrier mobility against carrier concentration in arsenic, phosphorus-, and boron-doped silicon," *IEEE Trans. Electron Dev.*, vol. 30, no. 7, pp. 764–769, 1983, doi: 10.1109/T-ED.1983.21207.

[35] P. Palestri et al., "A comparison of advanced transport models for the computation of the drain current in nanoscale nMOSFETs," *Solid. State. Electron.*, vol. 53, no. 12, pp. 1293–1302, 2009, doi: 10.1016/j.sse.2009.09.019.

[36] S. Datta, "Nanoscale device modeling: The Green's function method," *Superlattices Microstruct.*, vol. 28, no. 4, pp. 253–278, 2000, doi: 10.1006/spmi.2000.0920.

[37] A. R. Brown, A. Martinez, N. Seoane, and A. Asenov, "*Comparison of density gradient and NEGF for 3D simulation of a nanowire MOSFET*," in *2009 Spanish Conference on Electron Devices*, 2009, pp. 140–143, doi: 10.1109/SCED.2009.4800450.

[38] A. Martinez, N. Seoane, A. R. Brown, J. R. Barker, and A. Asenov, "3-D nonequilibrium green's function simulation of nonperturbative scattering from discrete dopants in the source and drain of a silicon nanowire transistor," *IEEE Trans. Nanotechnol.*, vol. 8, no. 5, pp. 603–610, 2009, doi: 10.1109/TNANO.2009.2020980.

[39] Y. Lee, D. Logoteta, N. Cavassilas, M. Lannoo, M. Luisier, and M. Bescond, "Quantum treatment of inelastic interactions for the modeling of nanowire field-effect transistors," *Materials (Basel).*, vol. 13, no. 1, p. 60, 2020, doi: 10.3390/ma13010060.

[40] M. G. Ancona, "Density-gradient theory: A macroscopic approach to quantum confinement and tunneling in semiconductor devices," *J. Comput. Electron.*, vol. 10, no. 1–2, pp. 65–97, 2011, doi: 10.1007/s10825-011-0356-9.

[41] X. An et al., "Design guideline of an ultra-thin body SOI MOSFET for low-power and high-performance applications," *Semicond. Sci. Technol.*, vol. 19, pp. 347–350, 2004, doi: 10.1088/0268-1242/19/3/009/meta.

[42] G. Paasch and H. Übensee, "A modified local density approximation: Electron density in inversion layers," *Phys. Status Solidi Basic Res.*, vol. 113, no. 1, pp. 165–178, 1982, doi: 10.1002/pssb.2221130116.

[43] M. G. Ancona and H. F. Tiersten, "Macroscopic physics of the silicon inversion layer," *Phys. Rev. B*, vol. 35, no. 15, pp. 7959–7965, 1987, doi: 10.1103/PhysRevB.35.7959.

[44] A. Wettstein et al., "Quantum device-simulation with the density-gradient model on unstructured grids," *IEEE Trans. Electron Devices*, vol. 48, no. 2, pp. 279–284, 2001, doi: 10.1109/16.902727.

[45] A. Wettstein, "*Quantum Effects in MOS Devices.*" ETH Zürich, 2000, doi: 10.3929/ethz-a-004003951.

[46] F. M. Bufler, *Full-Band Monte Carlo Simulation of Nanoscale Strained-Silicon MOSFETS.* Hartung-Gorre, 2003.

[47] H. Kosina et al., "Theory of the Monte Carlo method for semiconductor device simulation," *IEEE Trans. Electron Devices*, vol. 47, no. 10, pp. 1898–1908, 2000, doi: 10.1109/16.870569.

[48] C. Jungemann S. Keith and B. Meinerzhagen, "Full-Band Monte Carlo device simulation of a Si/SiGe-HBT with a realistic ge profile," *IEICE Trans. Electron.*, vol. E83, pp. 1228–1234, 2000.

[49] X. Shao and Z. Yu, "Nanoscale FinFET simulation: A quasi-3D quantum mechanical model using NEGF," *Solid. State. Electron.*, vol. 49, no. 8, pp. 1435–1445, 2005, doi: 10.1016/j.sse.2005.04.017.

[50] C. S. Smith, "Piezoresistance effect in germanium and silicon," *Phys. Rev.*, vol. 94, no. 1, pp. 42–49, 1954, doi: 10.1103/PhysRev.94.42.

[51] F. Payet, "Modélisation et Intégration de Transistors a Canal de Silicium contraint pour les noeuds Technologiques CMOS 45nm et en deçà," PhD Thesis, University of Aix-Marseille, 2005.

[52] T. Guillaume, "Influence des contraintes mécaniques non-intentionnelles sur les performances des transistors MOS à canal ultra-court," PhD Thesis, Grenoble Alpes University, 2005.

[53] J. C. Maxwell, *A Treatise on Electricity and Magnetism.* Dover Publications, 1873.

[54] M. Lundstrom, "Drift-diffusion and computational electronics-still going strong after 40 years!," *Simul. Semicond. Process. Devices Conf.*, 2015, doi: 10.1109/SISPAD.2015.7292243.

[55] T. Grasser et al., "A review of hydrodynamic and energy-transport models for semiconductor device simulation," *Proc. IEEE*, vol. 91, pp. 251–274, 2003, doi: 10.1109/JPROC.2002.808150.

[56] G. Sasso, N. Rinaldi, G. Matz, and C. Jungemann, "*Analytical models of effective DOS, saturation velocity and high-field mobility for SiGe HBTs numerical simulation,*" in *2010 International Conference on Simulation of Semiconductor Processes and Devices*, 2010, pp. 279–282, doi: 10.1109/SISPAD.2010.5604505.

[57] B. H. Hong et al., "Possibility of transport through a single acceptor in a gate-all-around silicon nanowire PMOSFET," *IEEE Trans. Nanotechnol.*, vol. 8, no. 6, pp. 713–717, 2009, doi: 10.1109/TNANO.2009.2021844.

[58] C. Jungemann et al., "Stable discretization of the Boltzmann equation based on spherical harmonics, box integration, and a maximum entropy dissipation principle," *J. Appl. Phys.*, vol. 100, no. 2, p. 024502, 2006, doi: 10.1063/1.2212207.

[59] T. Grasser, R. Kosik, C. Jungemann, B. Meinerzhagen, H. Kosina, and S. Selberherr, "A non-parabolic six moments model for the simulation of sub-100 nm semiconductor devices," *J. Comput. Electron.*, vol. 3, no. 3–4, pp. 183–187, 2004, doi: 10.1007/s10825-004-7041-1.

[60] O. Penzin, L. Smith, A. Erlebach, and K. Lee, "Layer thickness and stress-dependent correction for InGaAs low-field mobility in TCAD applications," *IEEE Trans. Electron Devices*, vol. 62, no. 2, pp. 493–500, 2015.

[61] J. Colinge et al., "Quantum mechanical effects in trigate SOI MOSFETs," *IEEE Trans. Electron Devices*, vol. 53, no. 5, pp. 1131–1136, 2006, doi: 10.1109/TED.2006.871872.

[62] N. Pons et al., "*Density Gradient calibration for 2D quantum confinement: Tri-Gate SOI transistor application*," in *2013 International Conference on Simulation of Semiconductor Processes and Devices (SISPAD)*, 2013, pp. 184–187, doi: 10.1109/SISPAD.2013.6650605.

[63] G. Hiblot et al., "Impact of short-channel effects on velocity overshoot in MOSFET," 2015, doi: 10.1109/NEWCAS.2015.7182061.

[64] F. G. Pereira. Advanced numerical modeling applied to current prediction in ultimate CMOS devices, PhD Thesis, Universite Grenoble Alpes, 2016.

3 Stress Generation Techniques in CMOS Technology

Strain engineering is now considered one of the most important technology boosters of microelectronics among other technologies [1]. Strain can be introduced into devices unintentionally, for example, during the implantation and annealing steps. Nevertheless, taking advantage of the piezoresistive nature of silicon, the stresses are introduced voluntarily into the MOSFET transistors from the 90 nm node, in the interest of increasing the mobility of the charge carriers. Stress technologies are therefore today part of the boosters of microelectronics as well as the high-constant materials dielectric (high-k) or silicon on insulator. Their use tends to be generalized because the costs related to their introduction into processes are modest compared to performance gains made by other techniques [2]. The well-known "scaling" has been the driving force for improving the Power/Performance/Area. Nevertheless, the task has become more and more challenging over the years. Deformations introduced by stress/strain modify the band structure of silicon, in particular, causing degeneration lifts in the valence and conduction bands. The introduction of new transistor architectures has been necessary to meet the requirements both in terms of density and performance. Especially the use of strain as a performance booster has been widely discussed and is today mandatory in advanced technology nodes. In this chapter, we shall deal with the strain integration in advanced devices to boost and optimize the performance. In particular, dedicated attention is paid to the stress/strain profile mapping using simulation and compared with the measurement data.

In this chapter, the main strain integration techniques used in CMOS technology are presented briefly. These techniques can be sorted into two categories: the "locally introduced strain" techniques, in which the MOSFET channel is strained by the means of an external element, and "globally introduced strain techniques, in which the MOSFET channel material is intrinsically strained. Various methods have been envisaged for introducing strains into the silicon channel (see Figure 3.1). These processes are generally classified into two categories according to their mode of action, global or local, or according to the type of deformation they generate, biaxial or uniaxial [3]. A biaxial strain can be introduced by the silicon epitaxy on a relaxed SiGe substrate (Strain Relaxed Buffer technology). SiGe imposes its silicon lattice parameter, which has the effect of stretching the Si lattice in both directions of the plane and compressing it in the direction of growth. The biaxial processes are the first to have been developed because they are favorable both to the mobility of electrons and holes. For example, an increase in mobility of 110% for electrons and 45% for

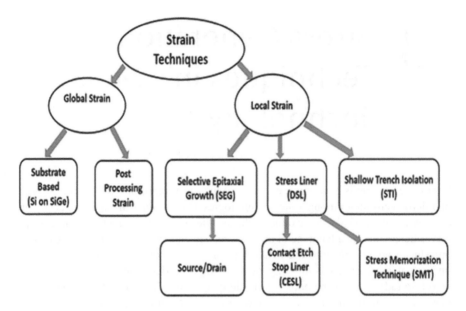

FIGURE 3.1   Various techniques are commonly used to introduce strain.

holes could be found for devices with a strained Si channel on a relaxed layer of SiGe ($x = 0.28$) [4, 5]. Note that a thin layer of Si on SiGe can also be transferred to SOI while maintaining its state of deformation [6]. This is called strained silicon on insulator (sSOI).

For uniaxial deformations, it is necessary to adapt the sign of the stress (tensile or compression) depending on the type of device. Indeed, a strain in tension is favorable to enhance the electron mobility in n-MOSFET while compressive deformation is favorable to enhance the mobility of holes for p-MOSFET. For this reason, uniaxial processes have encountered less interest in the beginning. Today, however, they are favored by the microelectronics industry, particularly because they rely on already more or less existing technologies. Various techniques are commonly used to introduce strain. Several methods can be envisaged to induce uniaxial deformation in a MOSFET:

Deposition of SiN stress-relieving film Contact Etch Stop Layer (CESL) enveloping the source, the gate, and the drain is performed. The film can induce tension or compression. The intensity of the stress variation depends on the deposition conditions and the thickness of the layer.

- *Growth of an alloy at the sources and drains,* SiGe, or SiC type. SiGe allows compression in the channel and is therefore used for p-MOSFETs. Conversely, for the carbon having a small lattice parameter concerning silicon, the SiC makes it possible to generate the tensile channel and can therefore be used in n-MOSFETs [7]. In both cases, the germanium and carbon content makes it possible to control the deformation.

- *Stress memorization technique (SMT)*. The general idea of this method is to deposit a SiN stressor film after the implantation step and before the annealing activation [8]. The film is then removed before the silicidation step. Finally, the state of stress of the transistor induced by the stressor film must be more or less preserved. The conservation of the deformation can be linked to the recrystallization of the gate [9].

Uniaxial processes are often local, that is, they are applied by introducing elements at the transistor scale. But, biaxial processes are generally global and applied on the wafer-scale (see Figure 3.2). The terms are therefore often confused although there are deformation processes that are uniaxial and global [10]. Uniaxial deformation has been adopted by Intel in the late 1990s [11]. Initially, SiGe was considered to reduce the resistance source and drain (better incorporation of dopants and better mobility) [12]. However, the measured drain currents were higher than the expected values. It has been concluded that SiGe deforms the crystal and thus modifies the

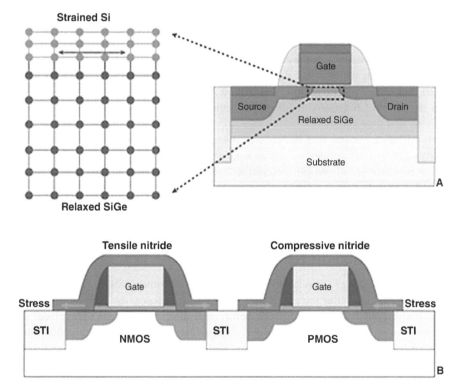

FIGURE 3.2 Representation of the main processes used to introduce strains into transistors. The principle of biaxial deformation is involved: when a thin layer of silicon is epitaxially on a relaxed substrate of SiGe, this stretches the Si lattice in both directions of the plane. Two uniaxial deformations involving the deposition of a SiN stressor film enveloping the source, gate, and drain can be used to introduce tension or compression.

mobility of the holes. Similarly, in the beginning, the layers of stopping SiN etching were not designed to induce strain. Nevertheless, it has been found that their use could alter the performance of devices by several percent [13]. Intel subsequently proposed a process to use SiGe source/drain on the same substrate for p-MOSFET and a SiN layer for n-MOSFET [14]. For a 90 nm technology, this method improved the saturation drain current by 25% (p-MOSFET) and 10% (n-MOSFET), as well as increase carrier mobility by 50% for holes and 20% for electrons.

3.1 STRESS/STRAIN MODELING

In this section, we briefly describe the theory of linear elasticity, particularly in the case of cubic structure materials. The general principle of simulation using finite element will then be presented and its implementation using Silvaco VICTORY Suite will be specified in 2D and 3D cases. In a solid material, a strain, that is, a change of shape occurs only if a system of forces is applied on its surfaces. These forces lead to the establishment of a system of strains (stress). The natural state in which no stress is applied is often considered the reference state.

A material is said to be in an elastic regime if the withdrawal of strains applied to it allows it to return exactly to its initial state. Beyond a certain limit, depending on the material considered, it may fracture or creep without being able to return to its original state. In which case, the elastic limit has been exceeded and the material enters a so-called plastic regime. Describing the elastic behavior of material consists of a large part of functional relations between the strains it undergoes and the deformations that they generate.

In all the definitions that follow, we consider the materials as homogeneous, that is to say, that their properties are independent of the coordinate system. We also define anisotropy. Anisotropic material will react differently depending on the direction of the stress that is applied. Conversely, in an isotropic situation, the effect of stress on the material is identical to whatever direction it is applied. If some external or internal forces are applied to the body, it deforms or changes its size.

3.1.1 LINEAR ELASTICITY THEORY

Hooke's law describes the relationship between stress σ and strain ε: For a rectangular material region, after the application of an external force, the body is deformed. This body represents the continuum before and after deformation. Therefore, the coordinates of all body points after deformation are continuous unambiguous functions of corresponding initial coordinates. The deformed body will have coordinates and components of the displacement vector as:

$$u_x = x' - x, u_y = y' - y, \left(u_z = z' - z, \text{ in case of 3D} \right) \tag{3.1}$$

where u_x, u_y, u_z are components of displacement vectors, or displacements.

In engineering science, the shear strain is defined as the change in angle between two originally orthogonal materials lines. The elastic theory of stress and strain uses the stress tensor $\underline{\underline{\sigma}}$ and the deformation tensor $\underline{\underline{\varepsilon}}$ of order 2 in a 3D space.

These tensors have nine components and can therefore be represented by 3×3 matrices in any reference system:

$$\underline{\underline{\sigma}} = \begin{pmatrix} \sigma_{11} & \sigma_{12} & \sigma_{13} \\ \sigma_{21} & \sigma_{22} & \sigma_{23} \\ \sigma_{31} & \sigma_{32} & \sigma_{33} \end{pmatrix} \tag{3.2}$$

$$\underline{\underline{\epsilon}} = \begin{pmatrix} \epsilon_{11} & \epsilon_{12} & \epsilon_{13} \\ \epsilon_{21} & \epsilon_{22} & \epsilon_{23} \\ \epsilon_{31} & \epsilon_{32} & \epsilon_{33} \end{pmatrix} \tag{3.3}$$

These two tensors are symmetrical, so that $\sigma_{ij} = \sigma_{ji}$ et $\epsilon_{ij} = \epsilon_{ji}$. By placing itself within the framework of the linear elasticity approximation, Hooke's law makes it possible to relate the stress tensor to that of the deformations according to the expression:

$$\sigma_{ij} = \Sigma_{kl} C_{ijkl} \epsilon_{kl} \ where \ (i,j,k,l) \in \{1;2;3\} \tag{3.4}$$

Moreover, the deformation tensor is connected to displacements according to:

$$\epsilon_{kl} = \frac{1}{2} \left(u_{k,l} + u_{l,k} \right) \tag{3.5}$$

where $u_{l,k}$ is shear deformation.

Hookes relation can also be written using the summation convention as:

$$\sigma_{ij} = C_{ijkl} \epsilon_{kl} \tag{3.6}$$

This relation involves the stiffness coefficients C_{ijkl} which form the tensor C of order 4, having 81 elements. However, the conditions of symmetry on the tensors of deformations and strain reduce the number of these components to 21 since:

$$C_{ijkl} = C_{jikl} = C_{ijlk} = C_{klij} \tag{3.7}$$

Equation (3.7) can be simplified further by using the Voigt notation which follows the rule of correspondence on the indices (i, j). Hookes law can then be expressed in matrix form. The tensors of the deformations and strain are represented as 6-component column vectors and the stiffness tensor is represented by a 6×6 matrices:

$$\begin{pmatrix} \sigma_{11} \\ \sigma_{22} \\ \sigma_{33} \\ \sigma_{23} \\ \sigma_{13} \\ \sigma_{12} \end{pmatrix} = \begin{pmatrix} C_{11} & C_{12} & C_{13} & C_{14} & C_{15} & C_{16} \\ C_{21} & C_{22} & C_{23} & C_{24} & C_{25} & C_{26} \\ C_{31} & C_{32} & C_{33} & C_{34} & C_{35} & C_{36} \\ C_{41} & C_{42} & C_{43} & C_{44} & C_{45} & C_{46} \\ C_{51} & C_{52} & C_{53} & C_{54} & C_{55} & C_{56} \\ C_{61} & C_{62} & C_{63} & C_{64} & C_{65} & C_{66} \end{pmatrix} \begin{pmatrix} \epsilon_{11} \\ \epsilon_{22} \\ \epsilon_{33} \\ \gamma_{23} = 2\epsilon_{23} \\ \gamma_{13} = 2\epsilon_{13} \\ \gamma_{12} = 2\epsilon_{12} \end{pmatrix} \tag{3.8}$$

3.1.2 CUBIC SYMMETRY MATERIALS: ANISOTROPIC CASE

In the case of silicon or germanium, cubic symmetry limits the number of components of the rigidity tensor with three independent coefficients. In the reference frame, the general expression (3.8) can be rewritten as follows:

$$
\begin{bmatrix}
\sigma_{x_0 x_0} \\
\sigma_{y_0 y_0} \\
\sigma_{z_0 z_0} \\
\sigma_{y_0 z_0} \\
\sigma_{x_0 z_0} \\
\sigma_{x_0 y_0}
\end{bmatrix}
=
\begin{bmatrix}
C_{11} & C_{12} & C_{12} & 0 & 0 & 0 \\
C_{12} & C_{11} & C_{12} & 0 & 0 & 0 \\
C_{12} & C_{12} & C_{11} & 0 & 0 & 0 \\
0 & 0 & 0 & C_{44} & 0 & 0 \\
0 & 0 & 0 & 0 & C_{44} & 0 \\
0 & 0 & 0 & 0 & 0 & C_{44}
\end{bmatrix}
\begin{bmatrix}
\epsilon_{x_0 x_0} \\
\epsilon_{y_0 y_0} \\
\epsilon_{z_0 z_0} \\
2\epsilon_{y_0 z_0} \\
2\epsilon_{x_0 z_0} \\
2\epsilon_{x_0 y_0}
\end{bmatrix}
\tag{3.9}
$$

To be able to enter the stiffness coefficients in simulation software, it is useful to present the tensor C in the form of 9 matrices of 3×3 as follows:

$$
\underline{\underline{C}} =
\left(
\begin{array}{ccc|ccc|ccc}
\begin{bmatrix} C_{11} & 0 & 0 \\ 0 & C_{12} & 0 \\ 0 & 0 & C_{12} \end{bmatrix}
&
\begin{bmatrix} 0 & C_{44} & 0 \\ C_{44} & 0 & 0 \\ 0 & 0 & 0 \end{bmatrix}
&
\begin{bmatrix} 0 & 0 & C_{44} \\ 0 & 0 & 0 \\ C_{44} & 0 & 0 \end{bmatrix} \\[20pt]
\begin{bmatrix} 0 & C_{44} & 0 \\ C_{44} & 0 & 0 \\ 0 & 0 & 0 \end{bmatrix}
&
\begin{bmatrix} C_{12} & 0 & 0 \\ 0 & C_{11} & 0 \\ 0 & 0 & C_{12} \end{bmatrix}
&
\begin{bmatrix} 0 & 0 & 0 \\ 0 & 0 & C_{44} \\ 0 & C_{44} & 0 \end{bmatrix} \\[20pt]
\begin{bmatrix} 0 & 0 & C_{44} \\ 0 & 0 & 0 \\ C_{44} & 0 & 0 \end{bmatrix}
&
\begin{bmatrix} 0 & 0 & 0 \\ 0 & 0 & C_{44} \\ 0 & C_{44} & 0 \end{bmatrix}
&
\begin{bmatrix} C_{12} & 0 & 0 \\ 0 & C_{12} & 0 \\ 0 & 0 & C_{11} \end{bmatrix}
\end{array}
\right)
\tag{3.10}
$$

Using Equation (3.9), the deformations can be expressed according to the strain:

$$
\begin{pmatrix}
\epsilon_{x_0 x_0} \\
\epsilon_{y_0 y_0} \\
\epsilon_{z_0 z_0} \\
2\epsilon_{y_0 z_0} \\
2\epsilon_{x_0 z_0} \\
2\epsilon_{x_0 y_0}
\end{pmatrix}
=
\frac{1}{D}
\begin{pmatrix}
C_{11}+C_{12} & -C_{12} & -C_{12} & 0 & 0 & 0 \\
-C_{12} & C_{11}+C_{12} & -C_{12} & 0 & 0 & 0 \\
-C_{12} & -C_{12} & C_{11}+C_{12} & 0 & 0 & 0 \\
0 & 0 & 0 & D/C_{44} & 0 & 0 \\
0 & 0 & 0 & 0 & D/C_{44} & 0 \\
0 & 0 & 0 & 0 & 0 & D/C_{44}
\end{pmatrix}
\begin{pmatrix}
\sigma_{x_0 x_0} \\
\sigma_{y_0 y_0} \\
\sigma_{z_0 z_0} \\
\sigma_{y_0 z_0} \\
\sigma_{x_0 z_0} \\
\sigma_{x_0 y_0}
\end{pmatrix}
\tag{3.11}
$$

with $D = (C_{11} - C_{12})(C_{11} + 2C_{12})$

3.1.3 CUBIC SYMMETRY MATERIALS: ISOTROPIC CASE

The presence of amorphous materials in transistors such as gate oxide leads us to also consider Hooke's law in an isotropic situation. In this case, the relationship can be expressed as follows:

$$
\begin{pmatrix} \sigma_{x_0 x_0} \\ \sigma_{y_0 y_0} \\ \sigma_{z_0 z_0} \\ \sigma_{y_0 z_0} \\ \sigma_{x_0 z_0} \\ \sigma_{x_0 y_0} \end{pmatrix} = \frac{E}{(1+v)(1-2v)} \begin{pmatrix} 1-v & v & v & 0 & 0 & 0 \\ v & 1-v & v & 0 & 0 & 0 \\ v & v & 1-v & 0 & 0 & 0 \\ 0 & 0 & 0 & \frac{(1-2v)}{2} & 0 & 0 \\ 0 & 0 & 0 & 0 & \frac{(1-2v)}{2} & 0 \\ 0 & 0 & 0 & 0 & 0 & \frac{(1-2v)}{2} \end{pmatrix} \begin{pmatrix} \epsilon_{x_0 x_0} \\ \epsilon_{y_0 y_0} \\ \epsilon_{z_0 z_0} \\ 2\epsilon_{y_0 z_0} \\ 2\epsilon_{x_0 z_0} \\ 2\epsilon_{x_0 y_0} \end{pmatrix}
$$

$$(3.12)$$

where E is Youngs modulus of the material and v it's Poisson's ratio.

To express Hooke's law (3.9) in the reference frame corresponding to the substrate orientation, it is necessary to make a basic change linked to the rotation of the coordinate system. Hookes law is then written in the reference frame as:

$$
\begin{pmatrix} \sigma_{x_1 x_1} \\ \sigma_{y_1 y_1} \\ \sigma_{z_1 z_1} \\ \sigma_{y_1 z_1} \\ \sigma_{x_1 z_1} \\ \sigma_{x_1 y_1} \end{pmatrix} = \begin{pmatrix} \frac{C_{11}+C_{12}+2C_{44}}{2} & \frac{C_{11}+C_{12}-2C_{44}}{2} & C_{12} & 0 & 0 & 0 \\ \frac{C_{11}+C_{12}-2C_{44}}{2} & \frac{C_{11}+C_{12}+2C_{44}}{2} & C_{12} & 0 & 0 & 0 \\ C_{12} & C_{12} & C_{11} & 0 & 0 & 0 \\ 0 & 0 & 0 & C_{44} & 0 & 0 \\ 0 & 0 & 0 & 0 & C_{44} & 0 \\ 0 & 0 & 0 & 0 & 0 & \frac{C_{11}-C_{12}}{2} \end{pmatrix} \begin{pmatrix} \epsilon_{x_1 x_1} \\ \epsilon_{y_1 y_1} \\ \epsilon_{z_1 z_1} \\ 2\epsilon_{y_1 z_1} \\ 2\epsilon_{x_1 z_1} \\ 2\epsilon_{x_1 y_1} \end{pmatrix}
$$

$$(3.13)$$

To simplify the coefficients, it is expressed as:

$$
\begin{pmatrix} \sigma_{x_1 x_1} \\ \sigma_{y_1 y_1} \\ \sigma_{z_1 z_1} \\ \sigma_{y_1 z_1} \\ \sigma_{x_1 z_1} \\ \sigma_{x_1 y_1} \end{pmatrix} = \begin{pmatrix} C'_{11} & C'_{12} & C'_{13} & 0 & 0 & 0 \\ C'_{12} & C'_{11} & C'_{13} & 0 & 0 & 0 \\ C'_{13} & C'_{13} & C'_{33} & 0 & 0 & 0 \\ 0 & 0 & 0 & C'_{44} & 0 & 0 \\ 0 & 0 & 0 & 0 & C'_{44} & 0 \\ 0 & 0 & 0 & 0 & 0 & C'_{66} \end{pmatrix} \begin{pmatrix} \epsilon_{x_1 x_1} \\ \epsilon_{y_1 y_1} \\ \epsilon_{z_1 z_1} \\ 2\epsilon_{y_1 z_1} \\ 2\epsilon_{x_1 z_1} \\ 2\epsilon_{x_1 y_1} \end{pmatrix}
$$

$$(3.14)$$

3.1.4 Constitutive Relations

The general relations between strain and stress tensors (Hooke's law) are as follows:

$$\sigma_{ij} = \Sigma_{k,l=1}^{3} C_{ijkl} \varepsilon_{kl} \tag{3.15}$$

$$\varepsilon_{ij} = \Sigma_{k,l=1}^{3} S_{ijkl} \sigma_{kl} \tag{3.16}$$

where C and S are the elasticity and compliance material coefficients, respectively. In the most general case, there are 81 coefficients. These coefficients C_{ijkl}, S_{ijkl} are tensors of rank 4. They could also depend on temperature or doping or both concentrations. It is convenient to use the matrix of elasticity C instead of the corresponding fourth rank tensor C_{ijkl}. The matrix C has 6×6 elements. Due to material symmetry, there are 21 different elements. In the case of isotropic materials, the elasticity matrix could be written using only two material parameters, Young's modulus (E) and Poisson's ratio (v). For anisotropic crystals with cubic symmetry (such as Si, Ge, GaAs) there are three independent elasticity constants: C_{11}, C_{12}, and C_{44}. For isotropic materials, these constants could be calculated by use of Young's modulus and Poisson's ratio:

$$C_{11} = \frac{E(1-v)}{(1+v)(1-2v)}, C_{12} = \frac{Ev}{(1+v)(1-2v)}, C_{44} = \frac{E}{2(1+v)} \tag{3.17}$$

For isotropic materials and anisotropic materials with cubic symmetry (when axes correspond to crystallographic directions), the elasticity matrix is following:

$$\begin{pmatrix} C_{11} & C_{12} & C_{12} & 0 & 0 & 0 \\ C_{12} & C_{11} & C_{12} & 0 & 0 & 0 \\ C_{12} & C_{12} & C_{11} & 0 & 0 & 0 \\ 0 & 0 & 0 & C_{44} & 0 & 0 \\ 0 & 0 & 0 & 0 & C_{44} & 0 \\ 0 & 0 & 0 & 0 & 0 & C_{44} \end{pmatrix} \tag{3.18}$$

C_{11}, C_{12}, and C_{44} are three independent elasticity constants.

The constitutive law Equation (3.15) can be written in matrix form by use of the relations between the strain and displacement functions as:

$$\varepsilon_{xx} = \frac{\partial u_x}{\partial x}, \varepsilon_{yy} = \frac{\partial u_y}{\partial y}, \varepsilon_{zz} = \frac{\partial u_z}{\partial z}$$

$$\varepsilon_{yz} = \frac{\partial u_z}{\partial y} + \frac{\partial u_y}{\partial z}, \varepsilon_{xz} = \frac{\partial u_z}{\partial x} + \frac{\partial u_x}{\partial z}, \varepsilon_{xy} = \frac{\partial u_y}{\partial x} + \frac{\partial u_x}{\partial y} \tag{3.19}$$

As the result, the constitutive relation (3.15) will be:

$$\sigma = C(\varepsilon - \varepsilon_0) + \sigma_0 \tag{3.20}$$

where $\varepsilon_0,$ σ_0 a are the initial strain and stress written in vector form for arbitrary matrix C. In this generic case, Equation (3.20) could be presented as a following system of equations:

$$
\begin{pmatrix} \sigma_{xx} \\ \sigma_{yy} \\ \sigma_{zz} \\ \sigma_{yz} \\ \sigma_{xz} \\ \sigma_{xy} \end{pmatrix} = C \times \begin{pmatrix} \varepsilon_{xx} - \varepsilon_{xx}^0 \\ \varepsilon_{yy} - \varepsilon_{yy}^0 \\ \varepsilon_{zz} - \varepsilon_{zz}^0 \\ \varepsilon_{yz} - \varepsilon_{yz}^0 \\ \varepsilon_{xz} - \varepsilon_{xy}^0 \\ \varepsilon_{xy} - \varepsilon_{xy}^0 \end{pmatrix} + \begin{pmatrix} \sigma_{xx}^0 \\ \sigma_{yy}^0 \\ \sigma_{zz}^0 \\ \sigma_{yz}^0 \\ \sigma_{xz}^0 \\ \sigma_{xy}^0 \end{pmatrix}
\tag{3.21}
$$

For the arbitrary axis orientation of anisotropic crystals of cubic symmetry, the elasticity matrix must be transformed by coming back to tensor notation.

For isotropic material, Equation (3.18) could be rewritten as a system of relationships for stress and strain components involving Young's modulus (E) and Poisson's ratio (v) (see also Equations (3.17, 3.18)) as:

$$
\sigma_{xx} = \frac{E(1-v)}{(1+v)(1-2v)}\left(\varepsilon_{xx} - \varepsilon_{xx}^0\right) + \frac{Ev}{(1+v)(1-2v)}\left(\varepsilon_{yy} - \varepsilon_{yy}^0\right)
$$
$$
+ \frac{Ev}{(1+v)(1-2v)}\left(\varepsilon_{zz} - \varepsilon_{zz}^0\right) + \sigma_{xx}^0
$$

$$
\sigma_{yy} = \frac{Ev}{(1+v)(1-2v)}\left(\varepsilon_{xx} - \varepsilon_{xx}^0\right) + \frac{E(1-v)}{(1+v)(1-2v)}\left(\varepsilon_{yy} - \varepsilon_{yy}^0\right)
$$
$$
+ \frac{Ev}{(1+v)(1-2v)}\left(\varepsilon_{zz} - \varepsilon_{zz}^0\right) + \sigma_{yy}^0
$$

$$
\sigma_{zz} = \frac{Ev}{(1+v)(1-2v)}\left(\varepsilon_{xx} - \varepsilon_{xx}^0\right) + \frac{Ev}{(1+v)(1-2v)}\left(\varepsilon_{yy} - \varepsilon_{yy}^0\right)
\tag{3.22}
$$
$$
+ \frac{E(1-v)}{(1+v)(1-2v)}\left(\varepsilon_{zz} - \varepsilon_{zz}^0\right) + \sigma_{zz}^0
$$

$$
\sigma_{yz} = \frac{Ev}{2(1+v)}\left(\varepsilon_{yz} - \varepsilon_{yz}^0\right) + \sigma_{yz}^0
$$

$$
\sigma_{xz} = \frac{Ev}{2(1+v)}\left(\varepsilon_{xz} - \varepsilon_{xz}^0\right) + \sigma_{xz}^0
$$

$$
\sigma_{xy} = \frac{Ev}{2(1+v)}\left(\varepsilon_{xy} - \varepsilon_{xy}^0\right) + \sigma_{xy}^0
$$

To obtain the elasticity matrix for anisotropic materials, the elasticity tensor rank-4 transformation has to be performed. After this transformation, in the case of cubic materials, the elasticity matrix Equation (3.18) for arbitrary axis orientations will be as follows:

$$C = \begin{pmatrix} C'_{11} & C'_{12} & C'_{13} & C'_{14} & C'_{15} & C'_{16} \\ C'_{12} & C'_{22} & C'_{23} & C'_{24} & C'_{25} & C'_{26} \\ C'_{13} & C'_{23} & C'_{33} & C'_{34} & C'_{35} & C'_{36} \\ C'_{14} & C'_{24} & C'_{34} & C'_{44} & C'_{45} & C'_{46} \\ C'_{15} & C'_{25} & C'_{35} & C'_{45} & C'_{55} & C'_{56} \\ C'_{16} & C'_{26} & C'_{36} & C'_{46} & C'_{56} & C'_{66} \end{pmatrix} \tag{3.23}$$

This matrix has 21 nonzero elements in a common case. The most important cases of wafer orientation are (100), (110), and (111). For these special cases of wafer orientation and arbitrary wafer flat rotation, this number of nonzero elements could be reduced down to 12. For anisotropic materials, a transformed elasticity matrix for specified wafer conditions in Equation (3.23) is used. Temperature and/or doping concentration dependencies of elasticity properties could be also used if specified.

3.1.5 Substrate Orientation

In 3D, the strain computed using Equation (3.24) is applied as a biaxial strain in the y- and z-directions.

$$u_x = x' - x, u_y = y' - y, \left(u_z = z' - z, \text{ in case of 3D} \right) \tag{3.24}$$

where u_x, u_y, u_z are components of displacement vectors or displacements. The initial strain for arbitrary substrate orientations can be calculated following reference [15] and is given here for some frequently used substrate orientations:

$$\varepsilon^0_{<100>} = \varepsilon \begin{pmatrix} 1 & 0 & 0 \\ 0 & 1 & 0 \\ 0 & 0 & -\dfrac{2C_{12}}{C_{12}} \end{pmatrix} \tag{3.25}$$

$$\varepsilon^0_{<100>} = \varepsilon \begin{pmatrix} \dfrac{2C_{44} - C_{12}}{C_{11} + C_{12} + 2C_{44}} & \dfrac{C_{11} + 2C_{12}}{C_{11} + C_{12} + 2C_{44}} & 0 \\ \dfrac{C_{11} + 2C_{12}}{C_{11} + C_{12} + 2C_{44}} & \dfrac{2C_{44} - C_{12}}{C_{11} + C_{12} + 2C_{44}} & 0 \\ 0 & 0 & 1 \end{pmatrix} \tag{3.26}$$

$$\varepsilon^0_{<111>} = \varepsilon \begin{pmatrix} \dfrac{4C_{44}}{C_{11}+2C_{12}+2C_{44}} & -\dfrac{C_{11}+2C_{12}}{C_{11}+2C_{12}+4C_{44}} & -\dfrac{C_{11}+2C_{12}}{C_{11}+2C_{12}+2C_{44}} \\[3mm] -\dfrac{C_{11}+2C_{12}}{C_{11}+2C_{12}+4C_{44}} & \dfrac{4C_{44}}{C_{11}+2C_{12}+2C_{44}} & -\dfrac{C_{11}+2C_{12}}{C_{11}+2C_{12}+4C_{44}} \\[3mm] -\dfrac{C_{11}+2C_{12}}{C_{11}+2C_{12}+4C_{44}} & -\dfrac{C_{11}+2C_{12}}{C_{11}+2C_{12}+4C_{44}} & \dfrac{4C_{44}}{C_{11}+2C_{12}+4C_{44}} \end{pmatrix} \quad (3.27)$$

The equilibrium assumption results in the following system of governing Equations (3.28) are solved for imposed boundary conditions depending on the specific problem. At these boundaries, the system of Equations (3.28) is satisfied with taking into account the corresponding stress components. In the case of the fixed boundary condition, Equation (3.29) is used.

$$\frac{\partial \sigma_{xx}}{\partial x} + \frac{\partial \sigma_{xy}}{\partial y} + \frac{\partial \sigma_{xz}}{\partial z} = 0$$

$$\frac{\partial \sigma_{xy}}{\partial x} + \frac{\partial \sigma_{yy}}{\partial y} + \frac{\partial \sigma_{yz}}{\partial z} = 0 \qquad (3.28)$$

$$\frac{\partial \sigma_{xz}}{\partial x} + \frac{\partial \sigma_{yz}}{\partial y} + \frac{\partial \sigma_{zz}}{\partial z} = 0$$

$$u_x = 0, \frac{\partial u_y}{\partial x} = 0, \frac{\partial u_z}{\partial x} = 0 \qquad (3.29)$$

The mechanical properties of the crystalline materials depend on orientation relative to the crystal lattice. This means that the correct values for analyzing two different designs in silicon may differ by up to 45%. However, perhaps, because of the perceived complexity of the subject, the common practices only do the isotropic calculations. As a result, many researchers oversimplify silicon elastic behavior and use inaccurate values for design and analysis. VictoryProcess provides accurate anisotropic effects through a tensorial transformation using substrate orientations as material axes. Elasticity is the relationship between stress and strain. Hookes law describes this relationship in terms of stiffness

$$\sigma' = C'.\varepsilon' \qquad (3.30)$$

It is seen that the primed directions denote the material coordinate system. In a general anisotropic material, a fourth rank tensor with 81 terms is required to describe the elasticity by relating the second rank tensors of stress and strain. Fortunately, in silicon, the combination of plane symmetry and the equivalence of the shear conditions allow us to specify the fourth rank tensor with a smaller number of independent components.

For cubic crystals, only three independent constants are required as follows:

$$\begin{Bmatrix} \sigma_1 \\ \sigma_2 \\ \sigma_3 \\ \sigma_4 \\ \sigma_5 \\ \sigma_6 \end{Bmatrix} = \begin{bmatrix} C_{11} & C_{12} & C_{12} & 0 & 0 & 0 \\ C_{12} & C_{11} & C_{12} & 0 & 0 & 0 \\ C_{12} & C_{12} & C_{11} & 0 & 0 & 0 \\ 0 & 0 & 0 & C_{44} & 0 & 0 \\ 0 & 0 & 0 & 0 & C_{44} & 0 \\ 0 & 0 & 0 & 0 & 0 & C_{44} \end{bmatrix} \begin{Bmatrix} \varepsilon_1 \\ \varepsilon_2 \\ \varepsilon_3 \\ \varepsilon_4 \\ \varepsilon_5 \\ \varepsilon_6 \end{Bmatrix} \tag{3.31}$$

For hexagonal crystals, nonzero and independent elastic stiffness constants are C_{11}, C_{12}, C_{13}, C_{33}, C_{44}, C_{66}. This matrix representation is given for a material coordinate system, which coincides with the default crystal orientation <100> as:

$$\begin{Bmatrix} \sigma_1 \\ \sigma_2 \\ \sigma_3 \\ \sigma_4 \\ \sigma_5 \\ \sigma_6 \end{Bmatrix} = \begin{bmatrix} C_{11} & C_{12} & C_{13} & 0 & 0 & 0 \\ C_{12} & C_{11} & C_{13} & 0 & 0 & 0 \\ C_{13} & C_{13} & C_{33} & 0 & 0 & 0 \\ 0 & 0 & 0 & C_{44} & 0 & 0 \\ 0 & 0 & 0 & 0 & C_{44} & 0 \\ 0 & 0 & 0 & 0 & 0 & C_{66} \end{bmatrix} \begin{Bmatrix} \varepsilon_1 \\ \varepsilon_2 \\ \varepsilon_3 \\ \varepsilon_4 \\ \varepsilon_5 \\ \varepsilon_6 \end{Bmatrix} \tag{3.32}$$

3.1.6 LATTICE MISMATCH STRAIN

Doping lattice mismatch is an intrinsic source for strain/stress. The top strained layer is grown depending on the strain required, and the lattice space is fixed by the lattice space in the relaxed region as shown in Figure 3.3.

The strain in the plane of the interface between the strained layer and the substrate is given by the lattice mismatch

$$\varepsilon = \frac{a_s - a_0}{a_0} \tag{3.33}$$

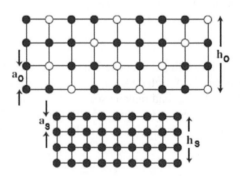

FIGURE 3.3 Lattice structure of unstrained Si and SiGe.

where a_o is the lattice constant of silicon and as is that of the substrate layer.

In 2D, the initial strain becomes

$$\varepsilon^0 = \varepsilon \begin{pmatrix} 1 & 0 \\ 0 & -\dfrac{C_{12}}{C_{11}} \end{pmatrix} \tag{3.34}$$

3.1.7 STRESS DUE TO THERMAL EXPANSION

One of the sources of strain/stress is the thermal expansion mismatch between material layers (e.g., between the substrate and overlying films). This source of strain is external and is given by:

$$\varepsilon_{ii}^0 = \int_{T_1}^{T_2} \alpha(T) dT \tag{3.35}$$

where α is the coefficient of thermal linear expansion and temperature changes from T_1 to T_2 Kelvin. The thermal linear expansion coefficient is a function of absolute temperature in the Kelvin scale. If this coefficient is a constant then Equation (3.30) will be as follows:

$$\varepsilon_{ii}^0 = \alpha \Delta T = \alpha (T_2 - T_1) \tag{3.36}$$

Note that the volume of material changes in each normal direction. Therefore, all initial shear strain components in Equations (3.20, 3.21) related to this thermal lattice mismatch will be equal to zero.

3.2 STRESS/STRAIN MAPPING IN SEMICONDUCTOR DEVICES

It is well known that strains or stresses occur during crystal growth and are frozen into crystalline ingots after cooling. These strains are responsible for several defects like dislocation generation. There is, however, there is a lack of a fast, non-destructive, and quantitative technique that can provide information on the residual strain and allow, in particular, spatially resolved high resolution (about 10 nm) as well as wafer-scale mapping. We introduce a new non-destructive approach, using TCAD simulation which addresses this strain mapping issue and demonstrates its feasibility by applying it to the spatially resolved evaluation of the stress/strain in semiconductor devices. The main objective of this section is to briefly review the experimental techniques currently in use for deformation analysis.

3.2.1 TECHNIQUES FOR STRAIN MEASUREMENTS

Since mobility enhancement using channel strain engineering has become one of the mainstream techniques for MOSFETs [16], the direct strain measurement technique plays an important role in understanding the mobility enhancement mechanism due to strain. According to ITRS reports [17], techniques using a transmission electron microscope (TEM) such as convergent beam electron diffraction and nanobeam diffraction (NBD) [18] are candidates for investigating methods of strain at a transistor level. Even using these techniques, direct strain measurement of the inversion layer of MOSFETs has not been achieved because their spatial resolution is up to 10 nm. Recently, strain mapping of strained MOSFET has been achieved with the TEM technique based on electron holography [19].

Advances in strain engineering have reinforced the need for precise deformation characterization methods. Deformation measurements are used at different stages of the design of CMOS devices. Commonly, three techniques of deformation measurements used in microelectronics are X-ray diffraction, Raman spectroscopy, and diffraction of backscattered electrons. X-ray diffraction is a non-destructive measurement method for obtaining information on the composition and the crystallographic structure of a material. For a given family of *hkl* planes, the diffracted intensity is maximal when Bragg's law is applied. When a crystal is illuminated by a monochromatic beam of light, it can be transmitted, reflected, absorbed, or diffused by the medium. The diffusion is predominantly elastic (Rayleigh scattering), nevertheless, for a tiny part of the incident beam, there may be an exchange of energy between the radiation and matter. This phenomenon was discovered in 1928 by physicist C. V. Raman [20]. The exchange of energy between radiation and matter implies a variation of frequency of the wave as well as the creation or annihilation of a phonon. It's about stokes shift when the frequency of the wave decreases (shift toward the red) and that there is the creation of a phonon.

Conversely, an anti-Stokes shift corresponds to a variation toward the high frequencies (toward the blue) and is associated with the annihilation of a phonon. This shift is independent of the excitation wavelength. It is a function of the type of chemical bonds involved and is therefore characteristic of the material studied. The Raman Effect can provide information on different properties of the medium: composition, structure, phase, crystalline orientation, but also on the presence of deformations. The existence of a strain in the material modifies the binding force constant and therefore the frequency of the Raman mode. The relationship between deformation and Raman frequency is quite complex because all the components of the tensor of deformation affect the frequency of Raman mode. However, in the case of biaxial stress or uniaxial, the relationship is simplified. The shift moves toward the low wave numbers if the stress is in tension and toward the high wavenumbers if it is in compression. Experimentally, strain within nanoscale-strained SiGe structures are investigated using a combination of reciprocal space maps collected using high-resolution X-ray diffraction exhibited distinct features corresponding to the SiGe and lattice deformation and were analyzed to quantify the state of stress within the films.

The technique is generally carried out in a scanning electron microscope (SEM) at an acceleration voltage of between 10 and 30 kV [21]. A polished sample is inserted into the device and is steeply inclined so that its surface forms an angle of 20° with the incident beam. Backscattered electrons, having undergone several elastic interactions with the matter, come statistically from all directions of space. All the planes of the analyzed area are then in Bragg's position. If the beam is stationary, an electron-backscatter diffraction (EBSD) emanates from the point of impact. A phosphor screen placed near (20 mm) of the sample intersects a portion of this sphere and allows to convert the electrons into the light to record a snapshot on a charge coupled device (CCD) camera. The figure of diffraction is formed of Kikuchi lines constituting a projection of the geometry of the crystal. Each band may be associated with the Miller indices of the diffracting plane. The relative position of these lines,, therefore, allows access to a certain amount of crystallographic information, especially on the size and the orientation of the grains, as well as on the crystalline phase.

TEM now has a range of different techniques to measure deformations with a resolution of a few tens of nanometers at 1 nm. Some are based on the study of interference fringes (high-resolution imaging, Moirés, holography), others on the analysis of diffraction patterns (convergent beam diffraction, nano diffraction), or the observation of contrast variations (dark field) [22]. The characteristic common to all these techniques is the need to prepare the sample in the form of an ultra-thin film transparent to electrons.

High-resolution transmission electron microscopy (HRTEM) is an imaging mode that can provide a representation of the crystallographic structure of a material at the atomic scale. This technique is particularly important for the study of the crystalline structure, especially for the analysis of the standard defects of grains and dislocations [23]. In practice, a sample of small thickness (<50 nm) is oriented in the zone axis and observed at high magnification (typically $\times 300,000$) using parallel illumination. The contrast comes from interference between the transmitted beam and the diffracted beams. It is called phase contrast in opposition to the amplitude contrast of conventional microscopy. However, the interpretation is not obvious as it depends on many parameters such as focus, astigmatism, aberrations of lenses. Simulations are often needed to determine the exact position of the atomic columns. Nevertheless, at the local level, it can be assumed that there is a constant relationship between the position of intensity extrema and atomic columns, which translates to the state of deformation. This is true if the sample is homogeneous in thickness and if the focus is properly adjusted.

The dark-field electron holography (DFEH) is a technique [24, 25] for generating deformation maps with a wide field of view, typically 500×500 nm^2, and a resolution of 5 nm. The principle is to create an interference between beams diffracted by the constrained region of interest and beams diffracted by a relaxed reference zone. It is necessary for this that these two zones are located close to each other (at a distance of less than 1 µm) and that they have the same crystallographic orientation. For this reason, the DFEH applies well to samples from the microelectronics, for which the silicon substrate serves as a reference.

Nanobeam electron diffraction (NBED) technique is relatively recent [26, 27] since it appeared in the early 2000s. The use of an almost parallel electron beam and typically less than 10 nm in size. Diffraction patterns of small diameter spots are recorded in series as the probe scans the sample. Variations in the position of the spots, compared to a reference snapshot taken in a non-deformed zone, can be connected to the deformation state in the probed zone.

3.3 SIMULATION OF DEFORMATION

In the following, we describe the methods used in the work to simulate deformations in thin films specifically in 2D and 3D cases. After introducing the linear elasticity theory, particularly in the case of cubic structure materials, the general principle of simulation using finite element has been presented and its implementation in the VICTORY Suite has been discussed in this section. In a solid material, a strain, that is, a change of shape occurs only if a system of forces is applied on its surfaces. These forces lead to the establishment of strain/stress. The natural state in which no stress is applied is often considered the reference state. As an example, stress components in a trigate FinFET are considered. Stress components in the fin height, width, and gate length direction in the device coordinate system are shown in Figure 3.4a. Also shown is the silicon crystal coordinate system (CCS). The schematic of a (110)/(110) FinFET on a (100) wafer is shown in Figure 3.4b. A material is said to be in an elastic regime if the withdrawal of stress applied allows the material to return exactly to its initial state. Beyond a certain limit, depending on the material considered, it may fracture or creep without being able to return to its original state. In which case, the elastic limit has been exceeded and the material enters a so-called plastic regime.

After describing the elastic behavior of the material, the functional relations between the strain it undergoes, and the deformations that they generate are discussed. In the definitions that follow, we consider the materials as homogeneous, that is to say, that their properties are independent of the coordinate system. We also define anisotropy as anisotropic material that will react differently depending on the direction of the stress that is applied. However, in an isotropic situation, the effect of stress on the material is identical to whatever direction it is applied. The deformations obtained by simulation were taken for the reference substrate (usually silicon) in its relaxed state.

Flexible devices can experience different types of deformations such as tension, compression, shear, bend, and torsion. As a general approach to deal with any combination of those deformation modes, we can apply prescribed motions at boundaries. The new deformation stress model is a very robust simulation method

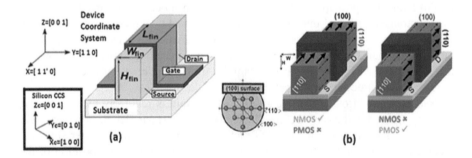

FIGURE 3.4 (a) Illustration of stress components in the fin height, width, and gate length direction in the device coordinate system. Also shown is the silicon crystal coordinate system (CCS). (b). Schematic of a (110)/(110) FinFET on a (100) wafer.

to analyze process-induced deformation stress as well as bending stress, and one can minimize its effect on the device characteristics through simulation without loss of cost and time.

Summarizing the elasticity theory described above the solution of the stress problem in 3D consists of the following steps in the VICTORY Suite:

1. The simulation domain and initial material structure are specified. Material properties such as Young's modulus and Poisson's ratio, or elasticity constants $C11$ $C12$ $C44$, are specified [28].
2. The system of equilibrium Equations (3.28) is solved in terms of displacement functions Equation (3.19) with the constitutive relations Equation (3.20) or (3.21) between stress and strain taken into consideration where the initial strain and stress written in vector form for arbitrary matrix C. For isotropic materials, elasticity properties Equation (3.22) are used. For anisotropic materials, the transformed elasticity matrix for specified wafer conditions Equation (3.23) is used. Temperature and/or doping concentration dependencies of elasticity properties could be also used if specified.
3. Initial values of strain/stress for sources described earlier are introduced. The equilibrium assumption results in the system of governing Equations (3.28) are solved for imposed boundary conditions depending on the specific problem. At these boundaries, the system of Equations (3.28) is satisfied with taking into account the corresponding stress components. In the case of the fixed boundary condition, Equation (3.29) is used.
4. The finite element method based on unstructured meshes (tetrahedral in 3D/triangular in 2D) is used for solving the system of Equations (3.28). The numerical results are the values of displacements in three directions at each mesh point. By the use of Equation (3.19), six components of strain are obtained. Finally, six components of stress are calculated by the use of Equation (3.21).

3.4   EPITAXIAL LAYERS: STRESS SIMULATION

The technique of epitaxial growth has been one of the cornerstones supporting the whole semiconductor industry since its emergence in the early 1960s. It is well known that SiGe follows a Stranski–Krastanov growth mode, which proceeds via the growth of bidimensional layers followed by the growth of 3D islands. Under this generic "Stranski–Krastanov" designation, several different behaviors can be identified. Moreover, the SiGe epitaxy has drawn considerable attention not only because of its great potential applications but also because it is regarded as a prototype system of heteroepitaxy since the two materials have the same crystallographic structure. Various models have been developed to predict the critical thickness for which the epitaxial strain layer can be grown. Van der Merwe produced a thermodynamic equilibrium model by minimizing the total energy of a system with the generation of a periodic array of dislocations. Positive signs of the stress values and the effective radius of curvature refer to tensile film stress, and negative values indicate compressive stress/strain, resulting in a tetragonal distortion of the lattice.

Silicon can be epitaxially grown as a thin layer on a relaxed SiGe substrate. Silicon epitaxial film will be in biaxial tension. The lattice parameter of SiGe alloys (denoted by a_{SiGe}) depends on the concentration of germanium. The higher the germanium concentration, the higher the lattice parameter. According to Vegard's law of the first order, this lattice parameter can be modeled according to the molar fraction of germanium x as:

$$a_{SiGe} = a_{Si}(1-x) + a_{Ge}x \tag{3.37}$$

with a_{Si} and a_{Ge} respective lattice parameters of silicon and germanium. It is important to specify that there are other accurate relationships, of the higher-order, more precise, which make it possible to calculate the lattice parameter of the SiGe alloys as a function of the x fraction of germanium [29, 30].

In practice, calculating the residual stress in the epitaxial layer requires the calculation of the deformation tensor associated with this layer, and then apply Hooke's law. The deformation of the silicon layer in the growth plane is equal to the relative difference in the value of its lattice parameter relative to the relaxed state:

$$\frac{\Delta a_{SiGe}}{a_{SiGe}} = \frac{a_{SiGe} - a_{Si}}{a_{Si}} \tag{3.38}$$

If the growth plane is the plane (001) of silicon, the longitudinal strain $\varepsilon_{//}$ in the plane of growth of the layers can be expressed in the system of crystallographic axes ([100], [010] and [001]) by:

$$\varepsilon_{//} = \varepsilon_{11} = \varepsilon_{22} = \frac{\Delta a_{SiGe}}{a_{SiGe}} \tag{3.39}$$

While the transverse deformation (denoted by $\varepsilon_{\perp}$) perpendicular to the plane of growth is of the form (result obtained by minimizing the elastic energy stored in the layer):

$$\varepsilon_{\perp} = \varepsilon_{33} = -2\frac{C_{12}^{SiGe}}{C_{11}^{SiGe}}\varepsilon_{//} \tag{3.40}$$

with C_{ij}^{SiGe} the elastic constants of SiGe material, expressed by:

$$C_{ij}^{SiGe} = (1-x)C_{ij}^{Si} + xC_{ij}^{Ge} \ (i = 1 \, at \, j = 1,2) \tag{3.41}$$

By applying Hooke's law and assuming that the component $\varepsilon_{\perp} = 0$, it is possible to go back to the stress tensor, whose $\sigma_{//}$ is the biaxial strain in-plane (001). The expression can be in the following analytic form:

$$\sigma_{//} = \sigma_{11} = \sigma_{22} = \varepsilon_{\perp}C_{12}^{SiGe} + \varepsilon_{//}\left(C_{11}^{SiGe} + C_{12}^{SiGe}\right) \tag{3.42}$$

By convention, a positive sign strain indicates stress in tension, while a negative sign corresponds to a compression.

To understand computer-aided device design, it is necessary to discuss the evolution of strain in the devices considered and the issues involved in their modeling. The global strain on the wafer level is mostly induced by the epitaxial growth of $Si_{1-x}Ge_x$ and Si layers. Because of the lattice parameter of $Si_{1-x}Ge_x$ ($0 < x < 1$) alloys vary between 0.5431 A for Si ($x = 0$) and 0.5657 A for Ge ($x = 1$), tensile strain is induced in a silicon layer epitaxially grown on top of the SiGe layer. The compressive strain is induced in the SiGe layer epitaxially grown on top of a Si layer. In this technology, the degree of strain is controlled by changing the content of Ge in the $Si_{1-x}Ge_x$ layer, or by changing the thickness of the strained Si layer. In both cases, the strain is in the plane of the layer, but this strain also produces a perpendicular the film thickness and the substrate thickness, respectively.

During the growth of epitaxial heterostructures, the lattice constant difference between the film and the substrate could induce epitaxial biaxial strain. By carefully selecting substrates and/or buffer layers, different strain states can be achieved in materials in thin-film form. This epitaxial strain is not only a unique approach to tune and manipulates their functionalities but also an interesting approach to stabilize phases that are not stable under ambient conditions. For instance, elastic strain engineering is one of the most interesting approaches to tune physical properties [31–33]. Epitaxial strain plays an important role in tuning functionalities. The lattice mismatch provides the source to the intrinsic lattice strain, which develops during nucleation and growth. The in-plane thermal expansion coefficient mismatch between the film and the substrate could induce thermal stress during the cool-down process. The local strain approach through using tensile and/or compressive strain nitride layer has been used to optimize n- and p-MOSFET devices on the same wafer independently by applying different levels of strain. More than 2 GPa of tensile stress and more than 2.5 GPa of compressive stress have been developed through controlling the growth conditions of Si_3N_4 layers [34].

Under appropriate growth conditions, good quality layers of crystalline $Si_{1-x}Ge_x$ alloys on Si substrates can be grown. If the SiGe thickness remains below a critical thickness, which depends on the alloy composition and the growth temperature, a pseudomorphic $Si_{1-x}Ge_x$ film can be grown without the introduction of extended defects. If the $Si_{1-x}Ge_x$ thickness exceeds the critical thickness, or the substrate is exposed to sufficiently high temperatures for a long time, at which the pseudomorphically grown layer is no longer thermodynamically stable, the lattice constant relaxes to its original value. This means that the strain in the $Si_{1-x}Ge_x$ layer will be relaxed and misfit dislocations will generate at the $Si/Si_{1-x}Ge_x$ interface. Thus, the $Si_{1-x}Ge_x/Si$ strained heterostructures are limited in thickness and stability.

3.5 SIMULATION CASE STUDIES

3.5.1 STRESS HISTORY MODEL

The ultimate goal of stress simulation is an accurate prediction of the stress distribution within the semiconductor from process modeling. The fabrication of integrated circuit devices requires a series of processing steps called the process flow. Process simulation involves modeling all essential steps in the process flow to obtain stress profiles. Process modeling is typically used as an input for device simulation from which the device's electrical characteristics are obtained. Most commercial processes and stress simulators consider the process flow, not as a physical process governed by a time-dependent model but rather a mathematical abstraction, where a new layer just appears on the top of the structure in case of deposition [35]. In these simulators, the stress equations are solved after a material layer or region has been already added to the structure and not during the physical process of deposition. This one-step approach cannot accurately predict stresses built in the deposited layer and adjacent areas. To obtain an accurate stress profile during these steps, the process should be considered as a series of deposition and relaxation steps to emulate mechanical quasi-equilibrium during the physical deposition process.

The device fabrication process involves several critical steps such as deposition of material layers (some with specific intrinsic stress), patterning, and etching of these layers, and also the heating and cooling processes. Hence, the stresses present in the device structure also change during the processing steps as each step deals with materials, etching, and heat. This simulation of stress evolution in the device has become essential for device reliability and performance analysis [36]. Commonly, one calculates stresses in the final device structure with different stressor regions by specifying values of intrinsic stresses which is known as the "one-step" model where stresses are calculated only once. So, this method may not calculate the precise stress amount devolved during the entire process steps. To calculate the process-induced stress precisely, we need to follow the "Stress History Model" or "Stress Evolution Model."

To verify the accuracy of the stress calculation using these two models, we have considered two test structures (half of a MOSFET) [34]. In the first test structure, a 40 nm thick nitride layer has been deposited on top of the MOSFET for which stress values are calculated. In the latter case, each time 8 nm thick nitride has been deposited five times to grow an equivalent 40 nm thick nitride layer. The stress in the later structure is calculated each time. The intrinsic stress of the nitride layer has been taken as +1 GPa for both cases. The stress profile (XX-component) for the above two cases have been shown in Figures 3.5 and 3.6, respectively. In both cases, the stress profile is different. It is observed that the stress profiles in flat areas of deposited layers, it is rather uniform and similar for both models. In the one-step model, it can be seen that the stress profile is nonuniform in the corner area of nitride and spacer which is different in the case of the stress evolution model. It can be seen that the stress distribution is highly non-uniform at the corner of the spacer to the top edge of the film. The stress value in the stress history model is high (1.4–2 times) in the case of the stress evolution model compared to that of the one-step model.

Figure 3.7 shows the dependency of stress components on the number of sub-layers. In this case, 20 sub-layers of 8 nm thick each has been deposited. It shows how stress gradually increases with an increase in the number of layers. So, it can be observed that the stress along XX and YY are 0.35 GPa and −0.28 GPa, respectively, for the stress evolution model which was around 0.24 GPa and −0.15 GPa for the one-step model. It indicates that the "stress evolution model" is a more suitable model for stress analysis.

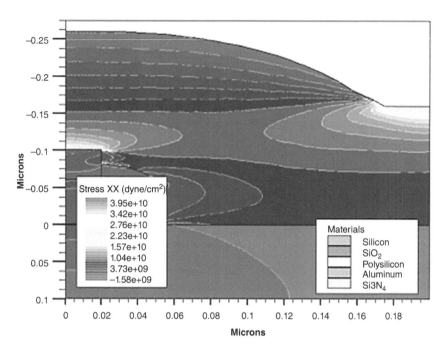

FIGURE 3.5 The Sxx distribution for the one-step model.

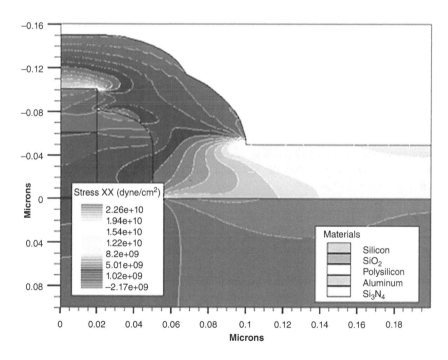

FIGURE 3.6 The Sxx distribution for stress evolution model with 5 sub-layers (Stress History Model).

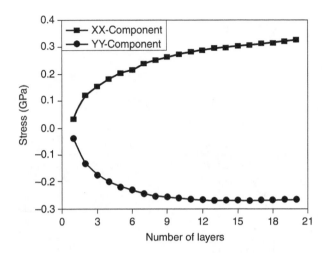

FIGURE 3.7 Average stress (both XX- and YY-components) variation with the number of sub-layers (8 nm thick) following the stress evolution model.

3.5.2 Lattice Mismatch Stress: Lattice Engineering

Stress develops in the film during the epitaxial deposition of a material on a substrate having a different lattice constant which also known as lattice engineering. The epitaxial layer gets strained as it tries to get aligned with the substrate below it. To illustrate the deposit of the SiGe layer, we simulated using the ATHENA [37]. Lattice mismatch stress is computed during the epitaxy or deposition stage to form a strained layer. The INIT material is considered a substrate by default. One can define any subsequent layer as a substrate by specifying the ISSUBSTRATE parameter into an EPITAXY or DEPOSIT statement. The structure showing the epitaxial growth of SiGe on Si substrate to form a heterostructure is shown in Figure 3.8. The stress developed in the above heterostructure is shown in Figure 3.9. It can be observed that compressive stress (negative values) is developed in the SiGe layer as the lattice constant of SiGe is higher than the Si lattice constant and dependent on the Ge concentration (Vegard's law).

The compressive stress is maximum in the region having the highest *Ge* concentration and then decreases accordingly. The stress is minimum in the middle of the SiGe layer where the *Ge* concentration is minimum. However, the quality of the interface between the SiGe layer and the gate oxide is of utmost concern as the carrier mobility is greatly affected by the high interface state density. Hence, to reduce the defect density at the interface, a thin silicon cap layer is inserted between the strained-SiGe channel and the gate oxide layer. During the CMOS fabrication process, silicon nitride is generally deposited over the oxide layer for isolation purposes. Figure 3.10 shows the stress generated in the oxide layer.

In the following, we consider the simulation of stresses in a MOSFET structure embedded in the SiGe layer. The simulated process includes the epitaxial growth of thin compressive SiGe layers. The most important step of this test process is the

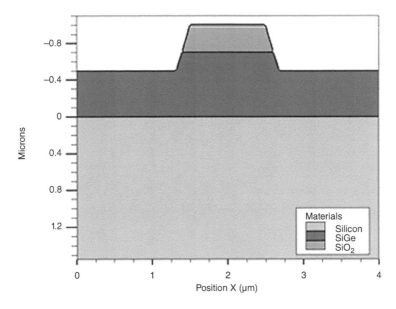

FIGURE 3.8 Epitaxial growth of SiGe on Si.

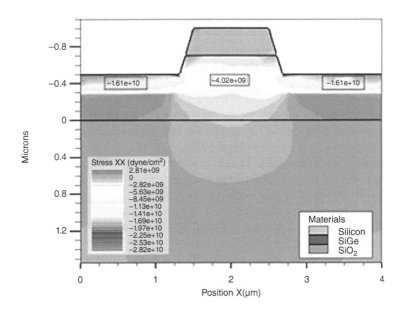

FIGURE 3.9 Stress in the SiGe layer grown on silicon due to lattice mismatch.

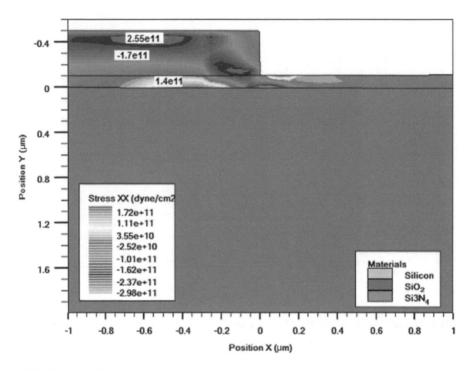

FIGURE 3.10   Stress distribution after oxidation.

etching of the source/drain areas because it creates free surfaces on the sides of the buried SiGe layer. This step results in elastic expansion of the buried layer, reducing the compressive stress inside the layer and generating tensile stress in silicon above, that is, under the gate. This enhanced tensile stress affects carrier transport and effectively improves device characteristics. The stresses in the whole structure are calculated before and after the S/D etch. Germanium concentration dependence of the stress XX component is shown in Figure 3.11.

Uniaxially tensile strained-Si layer is possible through the use of the new reverse embedded-SiGe-strained-Si processing technique which can significantly improve the n-FET performance. In this technique, a compressively strained and buried SiGe layer induces tensile strain in the overlying silicon channel. Though significant research has been carried out, the effect of layer thickness and optimization of the stress transfer has not been explicitly reported yet [38]. Stress transfer efficiency can be determined as follows: initially, a MOSFET structure is developed with epitaxially grown compressively strained SiGe layers on the silicon substrate. Then etching is done in the source/drain areas which creates a lateral free surface for the buried compressed SiGe layer to relax, thus decreasing the compressive stress in the SiGe layer and transferring tensile stress to the above silicon layer as shown in Figure 3.11. Germanium concentration dependence of the stress XX component is shown in Figure 3.12. In the device structure, before and after etching (see Figure 3.11) the resulting stress (cutline plot) in the SiGe and Si layers is shown in Figure 3.13.

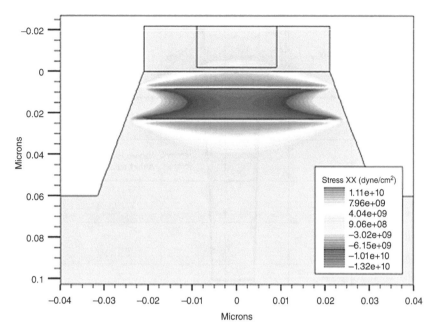

FIGURE 3.11 MOSFET structure showing the epitaxial SiGe and Si layers after etching.

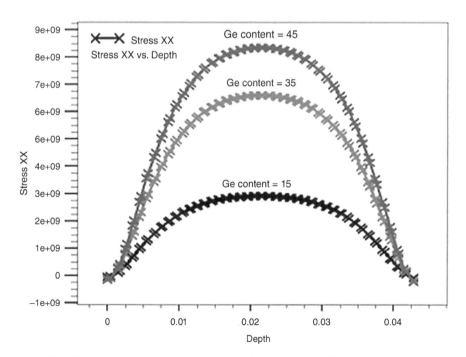

FIGURE 3.12 Germanium concentration dependence of stress XX component.

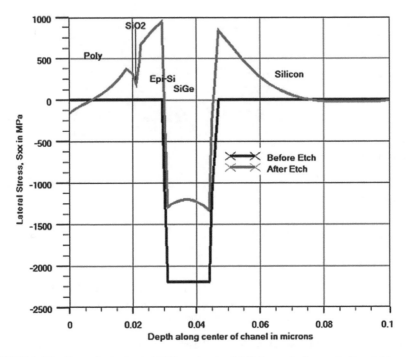

FIGURE 3.13   Stress in the buried SiGe and epitaxial Si layers before and after etching.

The induced tensile stress in the epitaxial Si layer is responsible for the drive current enhancement of n-MOSFETs.

In this case study, we also consider Si/SiGe and Si/SiGeC epitaxially deposited heterostructures. These structures may be used in the design of new generation transistors such as multichannel devices with bandgap grading. The introduction of carbon into the SiGe layers allows independent control of the deformation and the concentration of Ge [39]. However, the formation of inconsistent clusters of SiC during high-temperature annealing limits the applications of SiGeC. This leads to a modification of the deformation state of the lattice due to the reduction of the rate of substitutional carbon in the SiGeC matrix. The lattice mismatch stress XX component in graded composition SiGeC films is shown in Figure 3.14.

3.5.3   STRAIN-ENGINEERED MOSFETS

Field-effect transistors have been evolving through the structure transition from planar to 3D as device scaling continues beyond the 22 nm node. As part of this transition, 3D field-effect transistors have been extensively investigated and developed because of the excellent electrostatic control. Strain engineering continues to play an important role in improving transistor mobility in such 3D transistor structures. Among several techniques for strain engineering in 3D devices, epitaxial growth of $Si_{1-x}Ge_x$ or $Si_{1-x}C_x$ on the source/drain (S/D) region is especially viable due to the ease

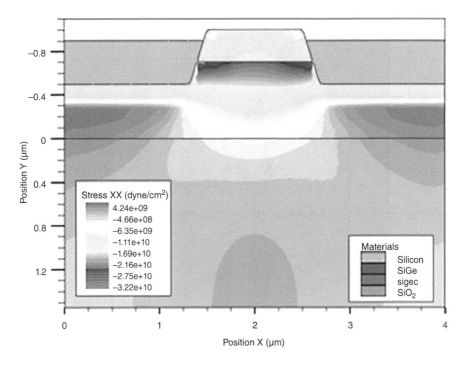

FIGURE 3.14 Lattice mismatch stress XX component in non-planar graded composition SiGeC films.

of integration and a reduction in contact resistance. However, $Si_{1-x}Ge_x$ or $Si_{1-x}C_x$ stressors tend to lose their effectiveness with a continued reduction in stressor volume as the transistor pitch scales. Quantitative nanoscale local strain profiling in embedded SiGe metal-oxide-semiconductor structures has been reported [40]. Experimentally, strain distribution in device structures with epitaxial $Si_{1-x}Ge_x$ stressors deposited around the source/drain region is obtained using the NBD technique.

We report herein on a systematic investigation of strain evolution in p-channel MOSFET structures induced by epitaxially grown SiGe layers using simulation [41]. This case study aims to study the characteristics of bandgap-engineered MOSFETs. An attempt has been made to provide a TCAD analysis of the bandgap engineered devices. For the analysis and design of any device, virtual fabrication of the devices is necessary. Through process simulation, virtual fabrication of the device is possible by specifying the process steps involved like deposition, lithography, etching, ion implantation, oxidation, and metallization. The heterostructure devices have been virtually fabricated using a novel CMOS process (changed from conventional CMOS process) technique in the ATHENA process simulator [37]. For MOSFET fabrication, the basic process steps are deposition, oxidation, ion implantation, lithography, and etching which are used to grow the field oxide, gate oxide, polysilicon, and metal layers of the MOSFET. The stress generated during the processing due to the lattice mismatch of Si and SiGe materials is considered in detail.

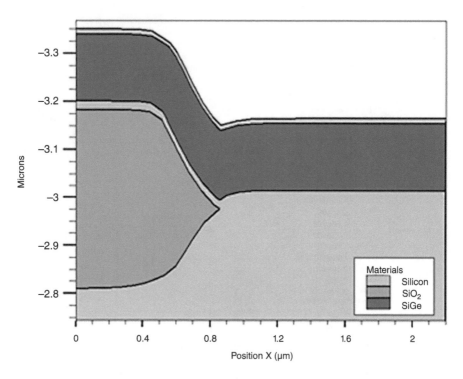

FIGURE 3.15   Deposition of the SiGe and Si layers.

In the initial step, the silicon substrate layer is grown on the Si wafer and doped with phosphorus impurity. Then the active area is defined on this substrate by the LOCOS process which involves the growth of the pad oxide and the nitride layer. After growing the field oxide, the pad oxide and the nitride layers are etched out from the active area on which the device has to be formed. In the next steps, the SiGe and the Si cap layers are deposited. The resultant structures from the above process flow are shown in Figure 3.15. Structure from process simulation after the deposition and patterning of the gate oxide and polysilicon is shown in Figure 3.16. The complete structure of the SiGe channel p-MOSFET from process simulation is shown in Figure 3.17.

After getting the complete device structure after metallization, the stress generated during the fabrication process in the device structure due to oxidation and deposition of the nitride layers has been studied and stress profiles at various stages are shown below. The stress along the X-axis and Y-axis are shown in Figures 3.18 and 3.19, respectively. The stress in the channel in both cases is found to be compressive. However, the stress developed along the X-axis (S_{XX}) is almost double that of the stress along the Y-axis (S_{YY}).

Following almost similar fabrication steps, strained-Si channel n-MOSFETs have been fabricated. The complete structure of the strained-Si channel n-MOSFET from process simulation is shown in Figure 3.20. Stress in the channel direction (S_{XX}) and stress in the vertical direction (S_{YY}) in the strained-Si n-MOSFETs are shown in Figures 3.21 and 3.22, respectively.

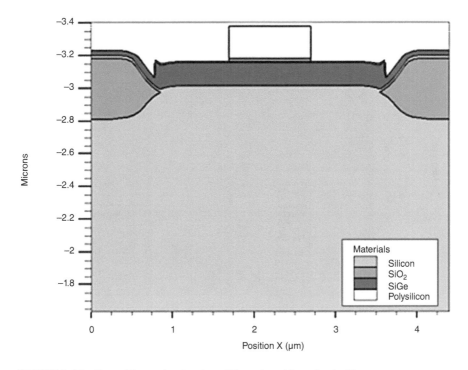

FIGURE 3.16 Deposition and patterning of the gate oxide and polysilicon.

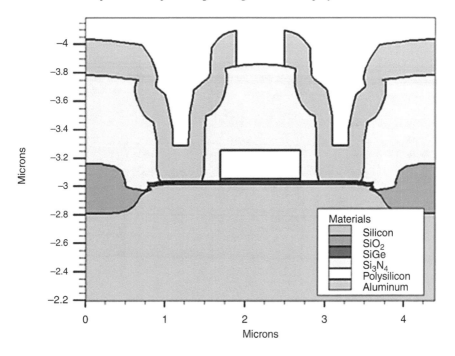

FIGURE 3.17 Structure of the SiGe channel p-MOSFET from process simulation.

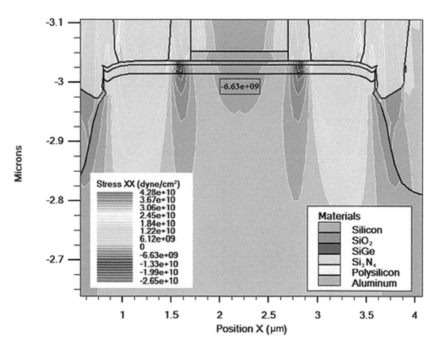

FIGURE 3.18   Stress distribution parallel to the channel direction in SiGe channel p-MOS-FET structure.

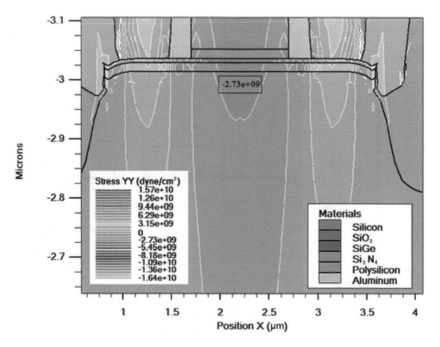

FIGURE 3.19   Stress distribution perpendicular to the channel direction in the SiGe channel p-MOSFET structure.

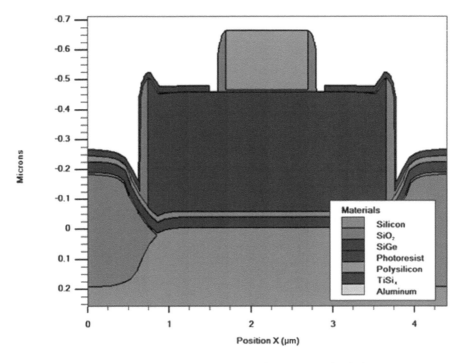

FIGURE 3.20 Structure of strained-Si n-MOSFET from process simulation.

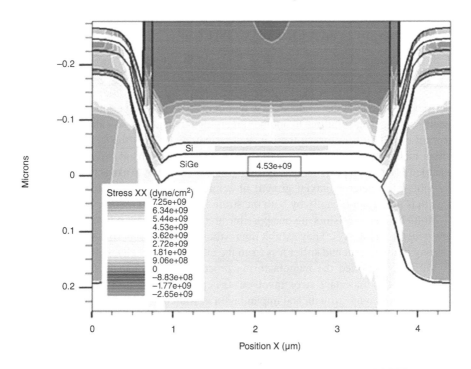

FIGURE 3.21 Stress in the channel direction (S_{XX}) in the strained-Si n-MOSFETs.

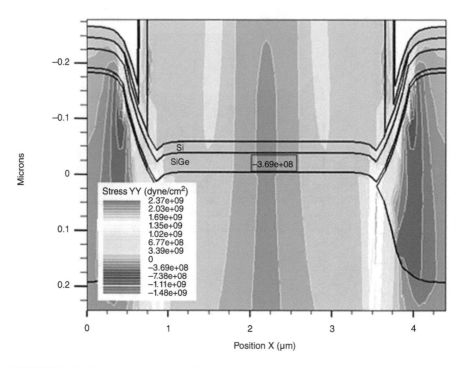

FIGURE 3.22   Stress in the vertical direction (S_{YY}) in the strained-Si n-MOSFETs.

3.5.4   SUPERLATTICE SiGe CHANNEL MOSFETs

In the early stages, epitaxial growth of SiGe on Si substrate attracted tremendous research interests due to its promising applications in the semiconductor integrated circuit industry [42]; then the interests have been soaring with the development of new devices based on the strained SiGe quantum wells [43], superlattices [44, 45] or quantum dots [46], solutions to another emerging demand, the so-called strain engineering which aims at tuning the physical properties through the strain induced by lattice mismatch between SiGe and Si [47]. On the other hand, SiGe/Si system is regarded as a prototype generic simple system that can serve to extract universal laws describing the heteroepitaxial growth of semiconductor nanostructures where the growth is tuned significantly by both the strain field and the kinetics.

Multichannel transistors are one possible architecture for MOSFET devices in the new generation [48, 49]. They exhibit interesting electrical properties with especially a strong drain current and suffer less from the effects of short channels. Denneulin et al. [50] have reported the manufacturing process of multichannel devices. Figure 3.23 shows the important steps involved: (a) epitaxial growth of Si/SiGe layer on SOI, (b) lithography, growth, and implantation of source/drain, (c) selective lithography of areas of SiGe, (d) formation of gate material HfO_2/TiN/polySi, and (e) TEM image of the multichannel device. Their manufacture is based first on the epitaxial growth of an array of alternating Si/SiGe layers on an SOI substrate as shown in Figure 3.23a. The structure is then etched to grow the source and drain areas

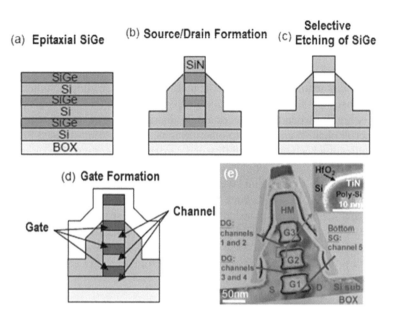

FIGURE 3.23 Main steps in the manufacturing process of multichannel devices. (a) Epitaxial growth of a Si/SiGe layer on SOI. (b) Lithography, then growth and implantation of source/drain. (c) Selective lithography of areas of SiGe. (d) Formation of gate material HfO_2/TiN/polySi. (e) TEM image TEM of a multichannel device [50].

(Figure 3.23b). Then, the SiGe layers are selectively etched to fill the voids with an HfO_2/TiN/polySi type gate material (Figures 3.23c, d). Figure 3.23e is a TEM image of a multichannel device.

It is difficult to grow a layer of SiGe on Si with a high concentration of germanium, necessary for lithography while maintaining good crystal quality. However, the addition of carbon in SiGe layers allows independent control of the germanium content and the level of deformation. Indeed, the introduction of a small amount of carbon in a substitutional position allows us to compensate for the deformation induced by a relatively large quantity of germanium. The rate compensation corresponds to a net strain equal to zero for an alloy of $Si_{(1-x-y)}Ge_xC_y$ when $\nu = x/y = 12$ [51]. The solubility of carbon in Si is very low (4.5×10^{17} atom.cm^{-3} at the melting point of silicon [52]. Nevertheless, it is possible to produce metastable SiGeC alloys containing 1–2% carbon with growth techniques such as molecular beam epitaxy or chemical vapor deposition. The incorporation rate of carbon in a substitutional position relative to the interstitial is nevertheless strongly dependent on growth conditions. Significant performance improvement of n/p MOSFETs has been simultaneously achieved in ultra-thin Si/Si0.6Ge0.4 multi quantum-well (MQW) channel structure [53]. The effects of SiGe and SiGe superlattice-like buried channels on n-FinFETs have also been investigated [54]. The device with a SiGe superlattice-like buried channel shows a 23% increase in electron mobility, a 29% increase in on-state current, and an I_{ON}/I_{OFF} ratio of $\sim 3 \times 10^6$. Silicon–germanium strained-layer superlattices for use as the buried channel in silicon p-MOSFETs have been proposed [55].

The aim is to maximize transconductance by increasing hole mobility in a sub-surface channel. Theoretical understanding of pseudomorphic SiGe on (100) Si as well as results of previous SiGe MOSFET designs suggests very high Ge mole fractions are required to obtain a significant performance increase. A superlattice avoids alloy scattering and potentially has higher hole mobility than SiGe alloys. Numerical modeling demonstrated the hole population in a Si/SiGe/Si heterostructure sub-surface channel is highly dependent on germanium distribution. The superlattice structure has a superior channel compared to other designs. Only Ge-rich strained SiGe layers offer significant advantages in mobility and carrier concentration. The unique aspect of the Si/SiGe superlattice sub-surface channel p-MOSFET is the presence of pseudomorphic Ge layers, albeit thin layers. This design is distinguished from other SiGe channel p-MOSFET designs which involved various profiles of $Si_{1-x}Ge_x$ alloys (in which x is not necessarily constant within the heterostructure). An experimental SiGe multichannel p-MOSFET [56] is shown in Figure 3.24. Figure 3.24c shows the schematic of the multichannel SiGe layers. Figure 3.24a shows a TEM image after SiGe etching. The simulated device structure before and after etching is shown in

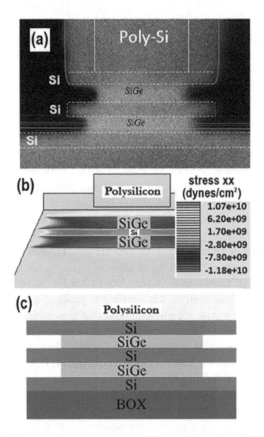

FIGURE 3.24 TEM image of an experimental SiGe multichannel p-MOSFET (a), simulated device structure before and after etching (b), and schematic of the multichannel SiGe layers (c) [56].

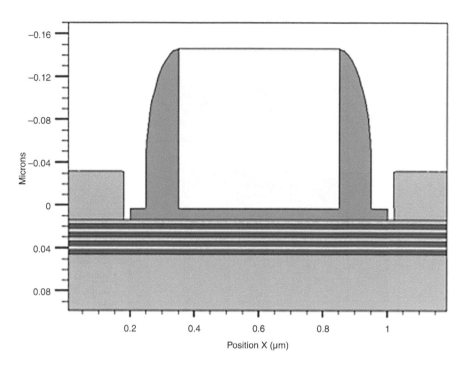

FIGURE 3.25   Structure of the 4-layer superlattice SiGe channel p-MOSFET from process simulation.

Figure 3.24b. It can be observed that compressive stress is developed in the SiGe layer as the lattice constant of SiGe is higher than the Si lattice constant and dependent on the Ge concentration. The induced tensile stress in the epitaxial Si layer is responsible for the drive current enhancement of the MOSFETs.

We have done a systematic investigation of the strain evolution in p-channel MOSFET structures induced by epitaxially grown SiGe layers using simulation (see Figure 3.25). This case study aims to study the characteristics of bandgap engineered superlattice p-MOSFETs. An attempt has been made to predict performance enhancement using TCAD analysis of the bandgap engineered superlattice devices. Through process simulation, virtual fabrication of the device is possible by specifying the process steps involved like the deposition, lithography, etching, ion implantation, oxidation, and metallization as discussed before. The heterostructure devices have been virtually fabricated using a novel CMOS process (changed from conventional CMOS process) technique in the ATHENA process simulator with four embedded SiGe channels separated by Si layer [37].

The quantitative nanoscale local strain profiling in embedded SiGe channels in the MOSFET is shown in Figure 3.26. As superlattice avoids alloy scattering, numerical modeling demonstrated a higher hole population in Si/SiGe/Si heterostructures. Thus, the superlattice MOSFETs show superior output characteristics (higher drain current) compared to single-channel conventional designs. As the number of embedded channels increases, the drain current increases (see Figure 3.27).

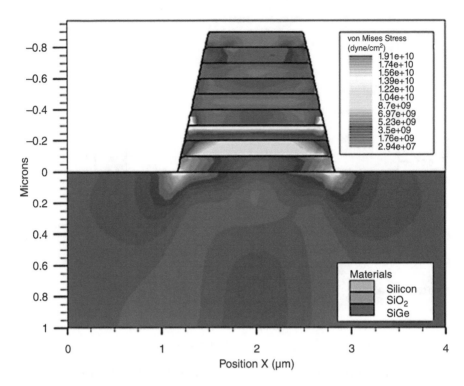

FIGURE 3.26   von Mises stress in superlattice SiGe layers grown on silicon due to lattice mismatch.

3.5.5 SUMMARY

Different techniques of stress introduction have been developed to boost the CMOS technologies. On the one hand, the local stressors are becoming less and less efficient as the dimensions are scaled-down. On the other hand, the introduction of the intrinsically strained channel has become necessary to achieve a high level of stress and therefore high performance. One of the main challenges of strain engineering is co-integration, as electrons and holes do not benefit from the same stress configuration.

By breaking the crystal symmetry, strain modifies the band structure of the Silicon. By inducing stress in the channel of MOSFETs, the charge carrier's mobility can be significantly increased. Consequently, there is now a need for a strain characterization at the nanoscale. A non-destructive new approach to strain mapping is necessary. The sensitivity and accuracy of the TCAD stress/strain simulation are evaluated. Then, the technique was applied to the study of different devices including superlattice MOSFETs that are used in the construction of new 3D architectures such as multichannel or multiwire transistors.

X-ray diffraction, μ-Raman, and CBED techniques have advantages and disadvantages that make them quite complementary. Raman is simple to implement, non-destructive and allows to quickly obtain information on the distribution of deformation provided that the mathematical model is well established for the sample considered.

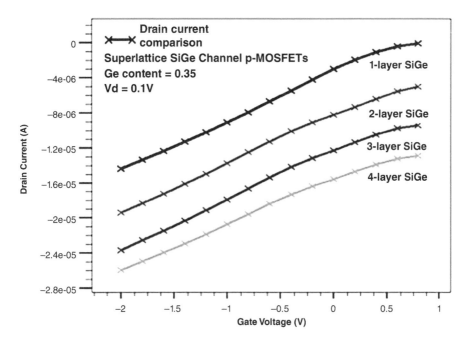

FIGURE 3.27 Output characteristics $(I_d\text{-}V_g)$ vs. the number of embedded channels for superlattice p-MOSFETs.

However, the spatial resolution is limited to several 100 nanometers. In contrast, the CBED has the advantage of offering nanoscale spatial resolution but the interpretation of the data is relatively complex. With its ability to perform spatially resolved measurements, TEM has become essential for stress characterization.

Alternatively, high-resolution images were also used to measure deformations by directly analyzing the position of the atomic columns. However, this approach suffers from a limited field of view and less appreciable sensitivity. In the early 2000s, NBED has emerged as a simple and effective technique for industrial applications. This technique involves the use of a parallel beam of nanometric size and allows us to obtain diffraction plates composed of spots of small diameters. 2D deformation is obtained directly from the variations in the position of the diffraction spots.

In 2008, a new technique called dark-field electronic holography has been developed based on the principle of holography, invented in 1948 by Dennis Gabor which allows through a process of interference and reconstruction, to access the wave phase; usually lost when recording a conventional snapshot.

We have examined the strain distribution in different MOSFET structures with epitaxial $Si_{1-x}Ge_x$ channels. Advanced stress/strain simulations allowed us to display how the strain induced by the stressors varies across the structure and, in turn, shows the channel strain. Besides, the strain was found very sensitive to such structural factors as the stressor thickness and fin width. The 2D strain variation data across the MOSFET structures obtained in this study are particularly beneficial for device design for MOSFETs, offering full information on strain variation across the structure.

REFERENCES

[1] C. K. Maiti, S. Chattopadhyay, and L. K. Bera, *Strained-Si Heterostructure Field-Effect Devices*. CRC Press (Taylor and Francis), USA, 2007.

[2] S. E. Thompson, "*Strained Si and the future direction of CMOS*," in *IEEE IDEAS Symp. Proc.*, 2005, pp. 14–16, 2005, doi: 10.1109/IWSOC.2005.99

[3] M. Reiche et al., "Strained silicon-on-insulator - fabrication and characterization," *ECS Trans.*, vol. 6, no. 4, pp. 339--344, 2019, doi: 10.1149/1.2728880.

[4] C. K. Maiti et al., "Hole mobility enhancement in strained-Si p-MOSFETs under high vertical fields," *Solid-State Electron.*, vol. 41, no. 12, pp. 1863–1869, 1997, doi: 10.1016/S0038-1101(97)00152-4.

[5] K. Rim et al., "*Strained Si CMOS (SS CMOS) technology: Opportunities and challenges*," in *Proc. Ultimate Integration of Silicon*, Munich, 2002, pp. 73–76.

[6] T. S. Drake et al., "Effect of rapid thermal annealing on strain in ultrathin strained silicon on insulator layers," *Appl. Phys. Lett.*, vol. 83, no. 5, pp. 875–877, 2003, doi: 10.1063/1.1598649.

[7] P. Verheyen et al., "*Demonstration of recessed SiGe S/D and inserted metal gate on HfO/ sub 2/ for high-performance pFETs.*," in *IEEE InternationalElectron Devices Meeting, 2005. IEDM Technical Digest.*, 2005, pp. 886–889.

[8] C. H. Chen et al., "*Stress memorization technique (SMT) by selectively strained-nitride capping for sub-65nm high-performance strained-Si device application*," in *IEEE VLSI Tech. Symp. Dig.*, 2004, pp. 56–57, doi: 10.1109/vlsit.2004.1345390.

[9] P. Morin et al., "A review of the mechanical stressors efficiency applied to the ultra-thin body and buried oxide fully depleted silicon on insulator technology," *Solid-State Electron.*, vol. 117, no. 3, pp. 100–116, 2016, doi: 10.1016/j.sse.2015.11.024.

[10] C. Himcinschi et al., "Strain relaxation in nanopatterned strained silicon round pillars," *Appl. Phys. Lett.*, vol. 90, no. 2, 2007, doi: 10.1063/1.2431476.

[11] M. Bohr, "The evolution of scaling from the homogeneous era to the heterogeneous era," 2011, doi: 10.1109/IEDM.2011.6131469.

[12] M. C. Ozturk, J. Liu, and H. Mo, "*Low resistivity nickel germanosilicide contacts to ultra-shallow Si1-xGex source/drain junctions for nanoscale CMOS*," in *IEEE IEDM Tech. Dig.*, Washington, 2003, pp. 497–500.

[13] S. Ito et al., "Mechanical stress effect of etch-stop nitride and its impact on deep submicron transistor design," *Tech. Dig. - Int. Electron Devices Meet.*, pp. 247–249, 2000, doi: 10.1109/IEDM.2000.904303.

[14] S. Thompson et al., "*A 90 nm logic technology featuring 50 nm strained silicon channel transistors, 7 layers of Cu interconnects, low k ILD, and 1 um2 SRAM cell*," in *Technical Digest - International Electron Devices Meeting*, 2002, pp. 61–64, doi: 10.1109/iedm.2002.1175779.

[15] J. M. Hinckley and J. Singh, "Influence of substrate composition and crystallographic orientation on the band structure of pseudomorphic Si-Ge alloy films," *Phys. Rev. B*, vol. 42, no. 6, pp. 3546–3566, 1990, doi: 10.1103/PhysRevB.42.3546.

[16] C. K. Maiti and T. K. Maiti, *Strain-Engineered MOSFETs*. CRC Press (Taylor and Francis), USA, 2012.

[17] ITRS Executive Summary - 2015 Edition., "International technology roadmap for semiconductors 2015." 2015.

[18] Y. Guo, G. Wang, C. Zhao, and J. Luo, "Simulation and characterization of stress in FinFETs using novel LKMC and nanobeam diffraction methods," *J. Semicond.*, vol. 36, no. 14121602, p. 86001, 2015, doi: 10.1088/1674-4926/36/8/086001.

[19] N. Nakanishi et al., "*Strain mapping technique for performance improvement of strained MOSFETs with scanning transmission electron microscopy*," in *2008 IEEE International Electron Devices Meeting*, 2008, pp. 1–4, doi: 10.1109/IEDM.2008.4796717.

[20] C. V. Raman and K. S. Krishnan, "A new type of secondary radiation," *Nature*, vol. 121, pp. 501–502, 1928.

[21] N. Gaillard et al., "Characterization of electrical and crystallographic properties of metal layers at deca-nanometer scale using Kelvin probe force microscope," *Microelectron. Eng.*, vol. 83, no. 11–12, pp. 2169–2174, 2006, doi: 10.1016/j.mee.2006.09.028.

[22] S. Tsuji, K. Tsujimoto, and H. Iwama, "Application of cross-sectional transmission electron microscopy to thin-film-transistor failure analysis," *IBM J. Res. Dev.*, vol. 42, no. 3.4, pp. 509–516, 1998, doi: 10.1147/rd.423.0509.

[23] K. S. T. Kiguchi N. Wakiya and N. Mizutani, *"HRTEM investigation of effect of various rare earth oxide dopants on epitaxial zirconia high-k gate dielectrics,"* in *Proc. Mat. Res. Soc.*, 2003, vol. 745, p. N9.7/T7.7.

[24] A. Durand, "Characterization and industrial control of local stress in microelectronics." PhD Thesis, Univ. of Grenoble, 2016.

[25] A. Beche et al., "Dark field electron holography for strain measurement," *Ultramicroscopy*, vol. 111, no. 3, pp. 227–238, 2011, doi: 10.1016/j.ultramic.2010.11.030.

[26] E. Jones et al., "Towards rapid nanoscale measurement of strain in III-nitride heterostructures," *Appl. Phys. Lett.*, vol. 103, no. 23, 2013, doi: 10.1063/1.4838617.

[27] A. Beche et al., "Improved precision in strain measurement using nanobeam electron diffraction," *Appl. Phys. Lett.*, vol. 95, no. 12, p. 123114, 2009, doi: 10.1063/1.3224886.

[28] W. A. Brantley, "Calculated elastic constants for stress problems associated with semiconductor devices," *J. Appl. Phys.*, vol. 44, no. 1, pp. 534–535, 1973, doi: 10.1063/1.1661935.

[29] S. Richard et al., "Energy-band structure in strained silicon: A 20-band k·p and Bir–Pikus Hamiltonian model," *J. Appl. Phys.*, vol. 94, no. 3, pp. 1795–1799, 2003, doi: 10.1063/1.1587004.

[30] M. V. Fischetti et al., "Band structure, deformation potentials, and carrier mobility in strained Si, Ge, and SiGe alloys," *J. Appl. Phys.*, vol. 80, no. 4, pp. 2234–2252, 1996, doi: 10.1063/1.363052.

[31] F. Isa et al., "Strain engineering in highly mismatched SiGe/Si heterostructures," *Mater. Sci. Semicond. Process.*, vol. 70, pp. 117–122, 2017, doi: 10.1016/j.mssp.2016.08.019.

[32] G. Tsutsui et al., "Strain engineering in functional materials," *AIP Adv.*, vol. 9, p. 030701, 2019, doi: 10.1063/1.5075637.

[33] F. Pezzoli, C. Deneke, and O. G. Schmidt, "Strain engineering of silicon-germanium (SiGe) micro- and nanostructures," in *Silicon-Germanium (SiGe) Nanostructures*, Editors: Y. Shiraki and N. Usami, Elsevier Ltd., 2011, pp. 247–295.

[34] S. Pidin et al., "A novel strain enhanced CMOS architecture using selectively deposited high tensile and high compressive silicon nitride films," in *IEEE IEDM Tech. Dig.*, 2004, pp. 213–216, doi: 10.1109/iedm.2004.1419112.

[35] K. V. Loiko et al., *"Multi-layer model for stressor film deposition,"* in *International Conference on Simulation of Semiconductor Processes and Devices, SISPAD*, 2006, pp. 123–126, doi: 10.1109/SISPAD.2006.282853.

[36] J. Lee and A. S. Mack, "Finite element simulation of a stress history during the manufacturing process of thin film stacks in VLSI structures," *IEEE Trans. Semicond. Manuf.*, vol. 11, no. 3, pp. 458–464, 1998, doi: 10.1109/66.705380.

[37] Silvaco International, Athena Users Manual, 2019.

[38] J. G. Fiorenza et al., "Detailed simulation study of a reverse embedded-SiGe strained-silicon MOSFET," *IEEE Trans. Electron Devices*, vol. 55, no. 2, pp. 640–648, 2008, doi: 10.1109/TED.2007.913084.

[39] C. K. Maiti and G. A. Armstrong, *Applications of Silicon-Germanium Heterostructure Devices*. Inst. of Physics Publishing, Bristol, 2001.

[40] W. Zhao, G. Duscher, G. Rozgonyi, M. A. Zikry, S. Chopra, and M. C. Ozturk, "Quantitative nanoscale local strain profiling in embedded SiGe metal-oxide-semiconductor structures," *Appl. Phys. Lett.*, vol. 90, no. 19, p. 191907, May 2007, doi: 10.1063/1.2738188.

[41] T. Numata et al., "Performance enhancement of partially- and fully-depleted strained-SOI MOSFETs and characterization of strained-Si device parameters," in *IEEE IEDM Tech. Dig.*, 2004, pp. 177–180, doi: 10.1109/IEDM.2004.1419100.

[42] J. M. Hartmann et al., "Mushroom-free selective epitaxial growth of Si, SiGe, and SiGe: B raised sources and drains," in *Solid-State Electronics*, 2013, vol. 83, pp. 10–17, doi: 10.1016/j.sse.2013.01.033.

[43] L. Witters et al., "*Strained Germanium quantum well pMOS FinFETs fabricated on in situ phosphorus-doped SiGe strain relaxed buffer layers using a replacement Fin process*," in *2013 IEEE International Electron Devices Meeting*, 2013, pp. 20.4.1–20.4.4, doi: 10.1109/IEDM.2013.6724669.

[44] D. Boudier et al., "Low frequency noise analysis on Si/SiGe superlattice I/O n-channel FinFETs," *Solid. State. Electron.*, vol. 168, 2020, doi: 10.1016/j.sse.2019.107732.

[45] G. Hellings et al., "*Si/SiGe superlattice I/O finFETs in a vertically-stacked Gate-All-Around horizontal Nanowire Technology*," in *2018 IEEE Symposium on VLSI Technology*, 2018, pp. 85–86, doi: 10.1109/VLSIT.2018.8510654.

[46] V. Le Thanh, "Fabrication of SiGe quantum dots: A new approach based on selective growth on chemically prepared H-passivated Si(100) surfaces," *Thin Solid Films*, vol. 321, pp. 98–105, 1998.

[47] Y.-C. Yeo, "*Enhancing CMOS transistor performance using lattice-mismatched materials in source/drain regions*," in *2006 International SiGe Technology and Device Meeting*, 2006, pp. 1–2, doi: 10.1109/ISTDM.2006.246557.

[48] N. Singh et al., "High-performance fully depleted silicon nanowire (diameter < 5 nm) gate-all-around CMOS devices," *IEEE Electron Device Lett.*, vol. 27, no. 5, pp. 383–386, 2006, doi: 10.1109/LED.2006.873381.

[49] E. Bernard et al., "Impact of the gate stack on the electrical performances of 3D multi-channel MOSFET (MCFET) on SOI," *Soild-State Electron.*, vol. 52, no. 9, pp. 1297–1302, 2008, doi: 10.1016/j.sse.2008.04.014.

[50] T. Denneulin, "Dark-field electron holography: A reliable technique for measuring strain in microelectronic devices," PhD Thesis, Université de Grenoble, 2012.

[51] D. De Salvador et al., "Lattice parameter of $Si_{1-x-y}Ge_xC_y$ alloys," *Phys. Rev. B*, vol. 61, no. 19, pp. 13005–13013, 2000, doi: 10.1103/PhysRevB.61.13005.

[52] F. Durand and J. C. Duby, "Carbon solubility in solid and liquid silicon—A review with reference to eutectic equilibrium," *J. Phase Equilibria*, vol. 20, pp. 61–63, 1999, doi: 10.1361/105497199770335956.

[53] L. K. Bera et al., "*Strained Si/SiGe multiple quantum well channels on bulk-si for high performance CMOS applications*," in *ULIS Conf. 2006*, Grenoble, 2006.

[54] Y. Li et al., "Improved electrical characteristics of bulk FinFETs with SiGe super-lattice-like buried channel," *IEEE Electron Device Lett.*, vol. 40, no. 2, pp. 181–184, 2019, doi: 10.1109/LED.2018.2890535.

[55] G. Abstreiter, "Physics and perspectives of Si/Ge heterostructures and superlattices," *Phys. Scr.*, vol. T49A, pp. 42–45, 1993, doi: 10.1088/0031-8949/1993/T49A/006.

[56] B. C. Paz, "Avaliação da influência da evolução das tecnologias de fabricação de nanofios transistores MOS sobre suas características elétricas," PhD Thesis, Centro Universitário FEI, 2018.

4 Electronic Properties of Engineered Substrates

The introduction of strain in silicon fundamentally changes the mechanical, electrical (band structure and mobility), and chemical (diffusion and activation) properties. Since the primary focus of this monograph is on strain-engineered devices, it is essential to understand the basics of engineering mechanics like stress, strain, and mechanical properties of the semiconductor involved. Within the elastic limit, the property of solid materials to deform under the application of an external force and to regain their original shape after the force is removed is referred to as elasticity. It is Hooke's law, which describes the elastic relationship between the mechanical strain and deformation that the material will undergo. The external force applied in a specified area is known as stress, while the amount of deformation is called the strain. In Chapter 3, the theory of stress, strain, and their interdependence has been discussed. In this chapter, we consider in particular the evolution of the piezoresistive effect and electrical transport properties in field-effect transistors involving different variables such as the geometry, temperature, and internal mechanical stress, to study the effects of the reduction at the extreme scaling limit of the channel and gate dimensions in the transistors. A model of the evolution of the piezoresistive coefficients has been presented. This model makes it possible to predict the variations of the piezoresistive coefficients with the cross section, viz., width, and thickness of the channel.

4.1 ENERGY GAP AND BAND STRUCTURE

The structure of crystalline silicon is a network face-centered cubic (FCC), with a diamond-like structure, and is illustrated in Figure 4.1. Each node in the network is composed of two atoms placed in positions (0, 0, 0) and (1/4, 1/4, 1/4). The basic cell (cell Wigner–Seitz) reciprocal lattice, commonly known as the Brillouin zone, is represented in Figure 4.2 and depends on the wave vector K. An electron in a solid is defined by its energy E and its wave function ψ linked by the Schrödinger equation as:

$$H\Psi = E\Psi \tag{4.1}$$

where H is the Hamiltonian of the system. In a periodic crystal lattice, the structure of the band is described in reciprocal space by the relations of dispersion $E(K)$.

4.1.1 SILICON CONDUCTION BAND

The minimum of the band of conduction is on the way Γ-X which corresponds to direction <100>. Silicon being a cubic crystal the directions <100>, <010>, <001>,

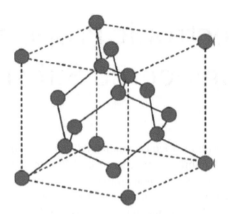

FIGURE 4.1   Crystalline structure of silicon.

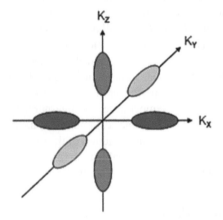

FIGURE 4.2   Ellipsoids of mass valleys Δ along with crystallographic principal directions.

<100>, <010>, and <001> are equivalent and gives us six equivalent minima, also called valleys Δ. Figure 4.2 shows the isoenergy surfaces around each of the six minimum conduction valleys. Six ellipsoidal surfaces are arranged according to the six directions equivalent to <100>. The wave functions are solutions of plane waves reflecting the decentralized nature of the particles. Relations dispersions are parabolic and written as:

$$E(k) = \frac{\hbar^2 k^2}{2m} \tag{4.2}$$

By breaking up this equation along the three axes, we obtain a relation of the type general equation of an ellipsoid:

$$E(k) = \frac{\hbar^2}{2m_0}\left(\frac{k_x^2}{m_x} + \frac{k_y^2}{m_y} + \frac{k_z^2}{m_z}\right) \tag{4.3}$$

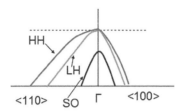

FIGURE 4.3   Valence bands in the directions <100> and <110>.

with m_x, m_y, and m_z effective masses, according to the wave vector k_x, k_y, and k_z, respectively. This equation describes in the space of k with constant energy an ellipsoid of mass.

4.1.2   SILICON VALENCE BAND

The valence band is filled up with holes; its maximum is centered in Γ point. Figure 4.3 shows the detail of the structure of bands in the directions <110> and <100>. Three bands coexist of (a) "heavy holes" or HH, (b) "holes light" or LH, and (c) "spin–orbit holes" or SO. We notice that the HH and LH bands are degenerated into their maximum but do not have the same ray of the curve and for the spin–orbit band is one, it is located at the lower part of the two others. Contrary to the bands of conduction, the valence bands are strongly anisotropic. In particular, heavy holes that occupy most of the valence band have a higher mass. Thus, the choice of the direction of transport becomes important for p-MOSFETs. The isoenergy surfaces are calculated using the k.p model [1, 2]. For simulation, however, a very simplified version implemented in MASTAR may be used [3, 4].

4.1.3   BAND STRUCTURE UNDER STRESS

Stress distorts semiconductor microstructures and results in changes in the band structure. In the deformation potential theory, the strains are considered to be relatively small. The change in energy of each carrier sub-valley, caused by the deformation of the lattice, is a linear function of the strain. By default (for silicon), simulation tools consider three sub-valleys for electrons (which are applied to three twofold sub-valleys in the conduction band) and two sub-valleys for holes (which are applied to heavy-hole and light-hole sub-valleys in the valence band). The number of carrier sub-valleys can be changed in the parameter file. We consider the Hamiltonian proposed by Bir and Pikus [5], which is finally combined with the Hamiltonian of Luttinger and Kohn [6], and that allows us to determine the effective masses and lifting of degeneration of the band.

In unstrained Si, the heavy-hole (HH) and the light-hole (LH) bands are degenerate at the Γ point as shown in Figure 4.4. The eigenstates at the Γ point split into two groups due to spin–orbit coupling and are classified by $J = 3/2$ and $J = 1/2$ and degenerated into HH ($J = 3/2$, $MJ = \pm 3/2$) and LH ($J = 3/2$, $MJ = \pm 1/2$) bands at the Γ point. The degeneracy between HH and LH bands is lifted due to the application of strain

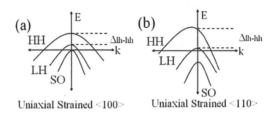

FIGURE 4.4 Simplified hole valance band structure for uniaxial strained in (a) <100> and (b) <110> direction.

on silicon crystal [7]. The Γ point at $k = 0$ has the full crystal symmetry. Uniaxial stress breaks the crystal symmetry by shortening (under compressive stress) and elongating (under tensile stress) and thus HH and LH bands are lifted. The HH and LH bands have negligible warping because they do not mix since the rotation symmetry is unchanged. Longitudinal uniaxial stress along <110> destroys the crystal symmetry more because the <100> axis has higher symmetry than the <110> axis. Due to the introduction of strain, the valence band warping for unstrained Si, Si under uniaxial compression takes place. The Si conduction band edges are located along the Δ valley and are six-fold degenerate. Due to the strain induced by the uniaxial stress, the six-fold degenerate Δ valley splits into two valleys, that is, Δ2 and Δ4 valleys, in either case. The splitting of energy and the ordering of the Δ2 and Δ4 valleys depend on the type and magnitude of the stress.

Because of its structure, silicon can be produced in wafers having as surface planes (100), used in industry, or also (100) and (111). The modification in the silicon substrate by the strain allows improvement in the performance of transistors, but several studies also show that changes through the use of different crystalline orientations of silicon [8, 9]. On these different planes are accessible different orientations, and the conventional direction is along the axis [110]. On the substrates (100) used, the devices oriented at 45° allow current conduction in the [100] direction, where the transport is different. It should also be noted that devices can be manufactured on substrates oriented at 45° from the beginning of the process so that the devices (defined by the lithography steps) are in the [100] direction intrinsically. The peculiarity of the use of orientation [100], for example, in trigate FinFET devices according to the direction [10], is that the available surfaces on the fins of the trigate are in the plane (100). It is therefore no longer possible to be in the presence of lateral surfaces (110) which improve the transport of the holes (in particular at small widths, when these surfaces predominate).

In the case of the transport of the holes, an increase of the effective mobility at the small widths W is due to the fins oriented according to (110). For the devices in the <100> direction, this gain in mobility is no longer possible since all the surfaces of the device are oriented in the plane (100). We observe lower mobility for holes. Irisawa et al. [11] reported that the <110> direction is the best configuration for a constrained Trigate device since the decrease in effective masses and the repopulation of the Δ2 valleys improve transport in this direction [9, 12]. The majority of the devices have been manufactured on silicon wafers having a crystalline plane (100), and with a channel oriented in the crystallographic direction <110>, conventionally used in the industry. Note that in this case, the orientation of the lateral faces of the trigate transistors is according to the plane (110). Destefanis et al. [13] have shown that these surfaces are denser in atoms of Si, are rougher and good control of the lithography is important, not to degrade the properties of the device. It is also possible to modify the shape of the active area using annealing under the H_2 atmosphere. This makes it possible to fill the dangling bonds on the surface of Si to finally create a section device elliptical or rounded. Several reports on these devices [14, 15] show that this removes the effects due to the confinement of the gate voltage in the corners, decreases the roughness of the surface, and allows a homogeneous control of the gate on the channel.

In contrast to the use of an sSOI substrate, allowing to bring a tensile strain directly into the manufactured transistors, creating a compressive strain can only be done through manufacturing processes. The most widely used technique is based on epitaxial source and drain SiGe regions with SiGe allowing to use of the larger lattice constant of Ge to strain the Si of the channel. Also, the use of the SiC epitaxy, with a lattice parameter less than Si, can be implemented to provide tensile stress [16]. The epitaxial source and drain areas can be selectively etched and the Ge content can be controlled through the decomposition of specific precursors [17], and thus change the level of strain.

The sectional view (see Figure 4.5) shows a 2D simulation of the carrier concentration calculated by the self-consistent resolution of the equations of Schrödinger–Poisson with 10 nm × 10 nm dimensions. It is observed that at a high gate voltage, the conduction is mainly at the SiO_2/Si interface. The dependence of mobility with the crystalline orientations of the surfaces of a multigate structure is very likely. It should be noted that these results are in agreement with the piezoresistive coefficients [19, 20] which show the dependence of the mobility variation with the strain σ and the piezoresistive coefficient π, according to the relation:

$$\frac{\Delta\mu}{\mu} = -\pi \times \sigma \qquad (4.4)$$

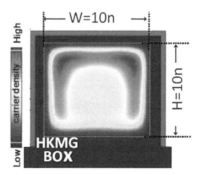

FIGURE 4.5 2D simulation carrier concentration profile calculated by the self-consistent solution of Schrödinger–Poisson equations [18].

4.2 PIEZORESISTIVITY MOBILITY MODELING

From a macroscopic point of view, the piezoresistive effect is characterized by the variation of electrical resistivity under the effect of mechanical stress applied to a material. In the case of an anisotropic crystal (e.g., silicon), the resistivity then forms a symmetric tensor of order 2. For sufficiently low-stress values (typically less than 200 MPa), the variation of resistivity is linearly related to the stress applied by a tensor of rank 4, called piezoresistivity tensor [21]. In a manner analogous to the properties of silicon elasticity, the piezoresistivity is also an anisotropic property. Due to the symmetry of stress tensors and resistivities resulting from symmetries of the silicon crystal (diamond structure) [22], three coefficients (π_{11}, π_{12}, and π_{44}) are sufficient for a complete description of the piezoresistivity of silicon instead of the 81 coefficients of the general case.

Piezoresistive coefficients are defined to describe the effect of stress on mobility in semiconductor materials. The piezoresistive coefficients extracted for uniaxial stress being lower than biaxial stress, the difference of gain in mobility is found qualitatively. In the piezoresistivity theory, one can assume a linear relationship between mobility and strain, for strain below 1 GPa. For higher strain (1.4 GPa), some deviations are observed [23]. Unlike solid material, the carriers in the inversion layer of a MOSFET are electrically confined (or geometrically for very thin films or very narrow FinFETs) and undergo additional interactions such as interface roughness Si/gate oxide or else Coulomb interactions due to charges in the gate. One can nevertheless always define piezoresistive coefficients which connect the conductivity of the channel at the applied stress. However, these coefficients will depend on the orientation of the inversion surface, and the symmetry of the coefficients. In the low-stress range, we consider that the mobility varies linearly with the stress, which has also been verified experimentally. This allows us to model piezoresistive coefficients for devices without initial stress. However, one needs to extend the definition of the piezoresistive coefficients for devices with a high initial tensile or compressive stress (>1 GPa).

Crystalline silicon is an anisotropic semiconductor material. For a crystal cubic symmetry (like silicon), the conductivity tensor is symmetrical and has six independent coefficients $\rho_{xy} = \rho_{yx}$; $\rho_{xz} = \rho_{zx}$; $\rho_{zy} = \rho_{yz}$. Besides, the diagonal elements of the tensor are constant and equal: $\rho_{xx} = \rho_{yy} = \rho_{zz} = \rho^0$, and the elements that are not diagonal are null:

$$\rho = \begin{pmatrix} \rho^0 & 0 & 0 \\ 0 & \rho^0 & 0 \\ 0 & 0 & \rho^0 \end{pmatrix} \tag{4.5}$$

When we apply a strain on the crystal, we change the symmetry of it and therefore its resistivity: it is the piezoresistive effect and we can describe this change in series as:

$$\frac{\Delta\rho_{ij}}{\rho^0} = \pi_{ijkl}\sigma_{kl} + \wedge_{ijklmn}\sigma_{kl}\sigma_{mn} + \dots \tag{4.6}$$

with π_{ijkl} and Λ_{ijklmn} the fourth- and sixth-order components of the piezoresistivity tensor, and $\sigma_{kl}\sigma_{mn}$ the components of the stress tensor. For low stress, the premier term dominates and the relation between the variation of resistivity and the strain is given by:

$$\frac{\Delta\rho_{ij}}{\rho^0} = \pi_{ijkl}\sigma_{kl} \tag{4.7}$$

The symmetry of the silicon allows simplifying the expression of the strain as:

$$\pi = \begin{pmatrix} \pi_{11} & \pi_{12} & \pi_{12} & 0 & 0 & 0 \\ \pi_{12} & \pi_{11} & \pi_{12} & 0 & 0 & 0 \\ \pi_{12} & \pi_{12} & \pi_{11} & 0 & 0 & 0 \\ 0 & 0 & 0 & \pi_{44} & 0 & 0 \\ 0 & 0 & 0 & 0 & \pi_{44} & 0 \\ 0 & 0 & 0 & 0 & 0 & \pi_{44} \end{pmatrix} \tag{4.8}$$

We thus find the resistivity with the stress applied expressed as:

$$\begin{pmatrix} \rho_1 \\ \rho_2 \\ \rho_3 \\ \rho_4 \\ \rho_5 \\ \rho_6 \end{pmatrix} = \begin{pmatrix} \rho^0 \\ \rho^0 \\ \rho^0 \\ 0 \\ 0 \\ 0 \end{pmatrix} + \rho^0 \begin{pmatrix} \pi_{11} & \pi_{12} & \pi_{12} & 0 & 0 & 0 \\ \pi_{12} & \pi_{11} & \pi_{12} & 0 & 0 & 0 \\ \pi_{12} & \pi_{12} & \pi_{11} & 0 & 0 & 0 \\ 0 & 0 & 0 & \pi_{44} & 0 & 0 \\ 0 & 0 & 0 & 0 & \pi_{44} & 0 \\ 0 & 0 & 0 & 0 & 0 & \pi_{44} \end{pmatrix} \begin{pmatrix} \sigma_1 \\ \sigma_2 \\ \sigma_3 \\ \sigma_4 \\ \sigma_5 \\ \sigma_6 \end{pmatrix} \tag{4.9}$$

with the indices 1, 2, 3, 4, 5, and 6 representing respectively, the directions xx, yy, zz, yz, xz, and xy according to vector notation. The electric vector field E can be expressed using the resistivity tensor and the current density vector, J (Ohm's law) as:

$$\begin{pmatrix} E_x \\ E_y \\ E_z \end{pmatrix} = \begin{pmatrix} \rho_1 & \rho_6 & \rho_5 \\ \rho_6 & \rho_2 & \rho_4 \\ \rho_5 & \rho_4 & \rho_3 \end{pmatrix} \begin{pmatrix} J_x \\ J_y \\ J_z \end{pmatrix} \tag{4.10}$$

To treat the case of channels oriented in any way concerning the crystallographic axes data, the same type of transformation laws are applied to the coefficients as for the elastic coefficients (change of reference). In general, the calculation of the longitudinal and transverse piezoresistive coefficients in the different crystallographic directions of n-doped silicon, shows a strong anisotropy of piezoresistive effects, especially in the silicon substrate planes (100) and (110). The variation of these coefficients in the plane (111) presents an isotropic character since the calculated coefficients have the same value whatever orientation.

In silicon as well as in the case of an inversion layer, the piezoresistive coefficients are generally well known [21, 24]. In contrast, few data are available for advanced CMOS devices [25]. The extraction of these parameters for short channels must take into account the impact of the access resistance. We can calculate in practice the mobility variation by using the coefficients of the piezoresistivity tensor when the applied stress is relatively low. For high values of strain of the order of GPa (typical case of current transistors), the use of the linear piezoresistive model is no longer appropriate. It then becomes necessary to apply non-linear models to relate the mobility variation to the strain [26, 27].

In the case of silicon, the piezoresistive approach only provides information on the interaction mechanism of carriers with phonons (strain-induced band structure modification). In contrast, in the case of MOSFET transistors where there is talk of the mobility of an inversion layer, and not a massive material, the piezoresistive approach alone is no longer adequate to justify the mobility in the channel, especially for transistors of small dimensions. Pelloux-Prayer et al. [28, 29] reported the variations in the piezoresistive coefficients induced by the reduction in the length and width of transistors. The addition of an important strain (of the order of the GPa) in a channel of a transistor leads to a significant variation of the piezoresistive coefficients. We are interested in the final resistivity in the direction of conduction. To calculate it, one projects the electric field on the normalized vector k which directs the direction of transport, $k = le_1 + me_2 + ne_3$ with $l^2 + m^2 + n^2 = 1$ (coordinates expressed in the main reference, see Figure 4.6):

$$\rho = \rho^0 \Big[\big(1 + \pi_{11}\sigma_1 + \pi_{12}(\sigma_2 + \sigma_3)\big)l^2 + \big(1 + \pi_{11}\sigma_2 + \pi_{12}(\sigma_1 + \sigma_3)\big)m^2$$
$$+ \big(1 + \pi_{11}\sigma_3 + \pi_{12}(\sigma_1 + \sigma_2)\big)n^2 + 2\pi_{44}(\sigma_6 lm + \sigma_5 ln + \sigma_4 mn) \Big] \qquad (4.11)$$

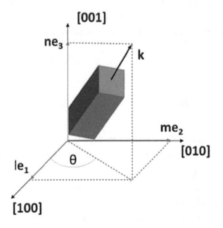

FIGURE 4.6 Channel orientation along with the vector k in the principal coordinate system.

We thus find a general expression of the piezoresistivity, whatever the direction of the transport, expressed in the main reference as:

$$\frac{\Delta\rho}{\rho} = \left(\pi_{11}\sigma_1 + \pi_{12}\left(\sigma_2 + \sigma_3\right)\right)l^2 + \left(\pi_{11}\sigma_2 + \pi_{12}\left(\sigma_1 + \sigma_3\right)\right)m^2$$
$$+ \left(\pi_{11}\sigma_3 + \pi_{12}\left(\sigma_1 + \sigma_2\right)\right)n^2 + 2\pi_{44}\left(\sigma_6 lm + \sigma_5 ln + \sigma_4 mn\right) \tag{4.12}$$

Generally, the transistors are oriented in the plane (100) of silicon. In the principal coordinate system, the orientation vector k of the channel will therefore always have its components following e_3 null ($n = 0$). Besides, the orientation vector k can be defined by an angle of rotation concerning the main coordinate system ($l = \cos(\theta)$, $m = \sin(\theta)$ see Figure 4.6), which simplifies the Equation (4.8) as:

$$\frac{\Delta\rho}{\rho} = \left(\pi_{11}\sigma_1 + \pi_{12}\left(\sigma_2 + \sigma_3\right)\right)\cos^2\left(\theta\right) + \left(\pi_{11}\sigma_2 + \pi_{12}\left(\sigma_1 + \sigma_3\right)\right)\sin^2\left(\theta\right)$$
$$+ 2\pi_{44}\sigma_6 \cos\left(\theta\right)\sin\left(\theta\right) \tag{4.13}$$

In the same way, if one expresses the stress initially in the principal reference, the stress in the reference (100) (turned by any angle) can be expressed as:

$$\sigma_1 = \sigma_1' \cos^2\left(\phi\right) + \sigma_2' \sin^2\left(\phi\right) - \sigma_6' \sin\left(2\phi\right)$$
$$\sigma_2 = \sigma_1' \sin^2\left(\phi\right) + \sigma_2' \cos^2\left(\phi\right) + \sigma_6' \sin\left(2\phi\right)$$
$$\sigma_3 = \sigma_3'$$
$$\sigma_4 = \sigma_4' \cos\left(\phi\right) + \sigma_5' \sin\left(\phi\right)$$
$$\sigma_5 = -\sigma_4' \sin\left(\phi\right) + \sigma_5' \cos\left(\phi\right)$$
$$\sigma_6 = \frac{\sin\left(2\phi\right)\left(\sigma_1' - \sigma_2'\right)}{2} + \cos^2\left(\phi\right)\sigma_6'$$

For uniaxial stress in the direction of the transistor channel, we have the following relation:

$$\frac{\Delta\rho}{\rho} = \frac{1}{2}\sigma_1'(\pi_{11}\left(1 + \cos^2\left(2\theta\right)\right) + \pi_{12}\left(1 - \cos^2\left(2\theta\right)\right) + \pi_{44}\sin^2\left(2\theta\right) \tag{4.14}$$

Likewise, for a uniaxial strain perpendicular to the transistor we have:

$$\frac{\Delta\rho}{\rho} = \frac{1}{2}\sigma_2'(\pi_{11}\left(1 - \cos^2\left(2\theta\right)\right) + \pi_{12}\left(1 + \cos^2\left(2\theta\right)\right) - \pi_{44}\sin^2\left(2\theta\right) \tag{4.15}$$

We can approximate the variation of the resistivity $\frac{\Delta\rho}{\rho}$ by $-\frac{\Delta\mu_i}{\mu_i}$ with i, the type of carriers, if the deformation creates a negligible amount of carriers. Indeed, we have by definition for a semiconductor:

$$\rho = \frac{1}{n \cdot q \cdot \mu_n + p \cdot q \cdot \mu_p} \tag{4.16}$$

By considering only the carriers i, the type of carriers, we obtain:

$$\frac{\Delta\rho}{\rho} = -\frac{\Delta i}{i} - \frac{\Delta\mu_i}{\mu_i} \tag{4.17}$$

The term i represents the carriers created by the stress and can be evaluated experimentally from the variation of the threshold voltage. It may be noted that for low stress, this term is negligible. For high strain, this term should be taken into consideration to estimate the piezoresistive coefficient.

The term $\frac{\Delta\mu_i}{\mu_i}$ represents the variation in mobility. This term is of our main interest as it determines the possible improvement in the performance of a MOSFET when the strain is applied to the channel. By convention, the mechanical stresses will be considered positive for tensile stress and negative for compressive stress. Concerning the piezoresistive coefficients (π_{ij}), this means that a tensile strain (therefore positive) will increase mobility for a negative piezoresistive coefficient, and compressive stress will increase mobility for a positive coefficient.

For channel orientation <100>, the piezoresistive relation for a longitudinal strain parallel to the direction is given by:

$$\frac{\Delta\mu}{\mu} = -\pi_L^{<100>}\sigma_L = -\pi_{11}\sigma_L \tag{4.18}$$

For a longitudinal strain perpendicular to the direction is given by:

$$\frac{\Delta\mu}{\mu} = -\pi_T^{<100>}\sigma_T = -\pi_{12}\sigma_T \tag{4.19}$$

For channel orientation <110>, the piezoresistive relation for a longitudinal strain parallel to the direction is given by:

$$\frac{\Delta\mu}{\mu} = -\pi_L^{<110>}\sigma_L = -\frac{\pi_{11} + \pi_{12} + \pi_{44}}{2}\sigma_L \tag{4.20}$$

and for strain perpendicular to the direction is given by:

$$\frac{\Delta\mu}{\mu} = -\pi_T^{<110>}\sigma_T = -\frac{\pi_{11} + \pi_{12} + \pi_{44}}{2}\sigma_T \tag{4.21}$$

4.2.1 ORIENTATION DEPENDENCE

For planar transistors, the piezoresistive measurements for a channel oriented in the <100> are interesting because they give direct access to the elementary components π_{11} and π_{12} of the piezoresistive tensor $\left(\pi_L^{<100>} = \pi_{11} \text{ et } \pi_T^{<100>} = \pi_{12}\right)$. Combined with the data in the <110> direction, we can thus have access to all elements of the tensor and therefore calculate all the cases in the plane (100). We know in particular that the sum of the coefficients $\pi_L + \pi_T$ is equal in the directions <100> and <110>:

$$\pi_L^{<110>} + \pi_T^{<110>} = \pi_L^{<100>} + \pi_T^{<100>} = \pi_{11} + \pi_{12} \qquad (4.22)$$

Besides, as explained before, the <100> FinFET have their surfaces side and top oriented in the plane (100). With these conduction surfaces being all similar, we can expect that the piezoresistive coefficients are independent of W_{top}.

4.2.2 PIEZORESISTIVE COEFFICIENTS FOR TRIGATE TRANSISTORS

Semi-analytical model of piezoresistive coefficients for trigate transistors has been derived from the mobility model and are given by [24]:

$$\frac{\Delta\mu_{TG}}{\mu_{TG}} = \frac{\Delta\mu_{top}}{\mu_{top}} \cdot \frac{1}{1 + \dfrac{2H}{W} \cdot \dfrac{\mu_{side}}{\mu_{top}}} + \frac{\Delta\mu_{side}}{\mu_{side}} \cdot \frac{1}{1 + \dfrac{W}{2H} \cdot \dfrac{\mu_{top}}{\mu_{side}}} \qquad (4.23)$$

$$\pi_{total} = \frac{\pi_{top}}{1 + \dfrac{2H_{NW}}{W_{top}} \cdot \dfrac{\mu_{side}}{\mu_{top}}} + \frac{\pi_{side}}{1 + \dfrac{W_{top}}{2H_{NW}} \cdot \dfrac{\mu_{top}}{\mu_{side}}} \qquad (4.24)$$

We can thus model the piezoresistive coefficient of a multigate device using conduction surface mobility, channel dimensions, and coefficients piezoresistive conduction surfaces. This expression empirically fit experimental data representing the coefficients π_{LT} as a function of W_{top}. In practice, μ_{side} and μ_{top} mobilities are estimated using the mobility model described before, and the piezoresistive coefficients π_{LT} is the adjustment parameters. Like before, the mobility of the upper surface is the mobility measured on wide devices, $W_{top} = 10\ \mu m$. The mobility of the lateral surfaces μ_{side} is given by the adjustment of the curves of μ_{eff} vs. W_{top} extrapolated to $W_{top} > 0$. The precise value of the mobility of lateral surfaces has only a slight impact on the accuracy of the piezoresistive parameters extracted up to $W_{top} = 1$ nm.

4.3 MOBILITY IN A TRANSISTOR

The mobility is derived from the movement of the charge carriers under the action of electric fields, processed statistically within the semiconductor. This treatment is done from the Schrödinger equation for the wave function of an electron (r,t) and the various potential disrupters that are likely to appear within a semiconductor component:

$$i\hbar \frac{\delta\Psi}{\delta t} = -\frac{\hbar^2}{2m_0} \nabla^2\Psi + \left[E_{C0}(r) + U_{C0}(r) + U_S(r,t) \right] \Psi(r,t) \qquad (4.25)$$

with $E_{C0}(r)$ the potential applied or undergone by the semiconductor, $U_{C0}(r)$ the potential crystalline, and $U_S(r,t)$ the dispersion potential (S for scattering).

The solution of this equation allows us to know more or less precisely (according to the methods used) transport conditions within the semiconductor component.

The concept of the effective mass of carriers (m) is derived from this approach in the diagram. It simplifies the study of the movement of carriers by grouping the concepts of crystalline potential and elementary mass of the electron ($m_0 = 9.1 \, Å \cdot 10^{-31}$ kg). The effective mass of carriers, expressed as a function of the elementary mass of electrons (i.e., $m^*_h = 0.19 \, m_0$), is inversely proportional to the radius of curvature of the energy band considered:

$$m^* = \hbar^2 \left[\frac{d^2 E}{dk^2} \right]^{-1}$$ (4.26)

We define the mobility of these carriers μ statistically by [30]:

$$\mu = \frac{q \, <\tau>}{m^*}$$ (4.27)

where q is the charge of the electron, τ is the average time between two collisions, and $<\tau>$ is the average on energy. This expression makes it possible to simply represent the concept of mobilization of the holders by involving the average free path $<\tau>$ of the holders, which is the average time between two collisions for a given carrier.

For silicon MOSFETs, the mechanisms of main interests are:

- interactions with phonons, which are the vibrations of the crystal lattice, surface roughness, which represents a disturbance of the conduction due to the physical proximity of the inversion layer and the non-ideal oxide-channel interface,
- Coulombian interactions, which represent the impact of the voltage present in the conduction channel if it is doped or in the oxide near the channel. These can be at the channel-oxide interface for D_{it} interface states, or in the oxide such as, for example, at the interface between the oxide SiO_2 and the oxide high-k [31]. For SiGe transistors, it is also necessary to consider the diffusion effect due to the presence of Ge in SiGe alloy [32].

The total mobility of the carriers in a transistor can be expressed, considering each interaction independently of each other, by Matthiessen law [33]:

$$\frac{1}{\mu_{tot}} = \sum_i \frac{1}{\mu_i} = \frac{1}{\mu_{ph}} + \frac{1}{\mu_{CS}} + \frac{1}{\mu_{SR}}$$ (4.28)

where μ_i represents the mobility limited by each mechanism: phonon μ_{ph}, the roughness of the surface μ_{SR}, and Coulomb interactions μ_{CS}. Phonon scattering is usually the predominant mechanism at room temperature for a mean electric field in the transistors. It is also the mechanism that is more sensitive to temperature. Experimentally, in the literature, limited mobility by the phonons in the transistors Si is proportional to T^α, with $\alpha \sim 1.75$ [34]. In theory, we distinguish interactions with phonons intervalley (inelastic phenomenon, the transition of the carriers of a sub-band of the valley in which it is present (Δ_2 or Δ_4 for example) to the sub-band of another valley) and interactions with intravalley phonons (elastic phenomenon, the transition of carriers from a sub-band to another sub-band of the same valley) [35]. These two types of interaction can be described analytically using the following equations [36]. For intravalley phonons, we have:

$$\frac{1}{\tau_{ac}^{1,2}} = \frac{n^{ac} m_d D_{ac} k_B T}{\hbar^3 \rho s_l} W_{1,2,} \tag{4.29}$$

$$W_{1,2} = \int \psi_1^2(z) \psi_2^2(z) dz \tag{4.30}$$

where m_d the density of state of the valley, D_{ac} the potential of acoustic deformation, ρ and s_l are respectively the mass density and the longitudinal velocity of the sound in the crystal. The term $W_{1,2}$ is the compound form factor of the envelope functions $\psi(z)$ sub-bands 1 and 2. In this expression, we find the variation of the mobility in the function of the inverse of the temperature ($1/T$).

For the interval phonons, we have:

$$\frac{1}{\tau_{iv}^{1,2}} = \frac{n^{iv} m_d^{(2)} D_{iv}^2}{\hbar \rho E_{iv}} \frac{1}{W_{1,2}} \left(N_{iv} + \frac{1}{2} \pm \frac{1}{2} \right) \times \frac{1 - f(E \mp E_{iv})}{1 - f(E)} U(E \mp E_{iv} - \Delta E_{1,2}) \tag{4.31}$$

with $m_d^{(2)}$ the effective mass of the state density of the final valley, D_{iv} the potential of deformation interval, E_{iv} the energy of the phonon interval, N_{iv}, and the number of occupation of Bose–Einstein inter-valley phonon for energy mode E_{iv}. The calculation of the mobility limited by the surface roughness is more complex and to intervene in the morphology of the interface via the parameters L_c (coherence length) and Δ (average roughness) [37, 38]. This contribution is generally accepted as being independent of temperature [34].

4.3.1 Mobility Models for Multigate Transistors

A simple way to describe transport in a multigate structure has been described in reference [39] and studied for FinFET transistors by Rudenko et al. [40]. In this description, the total mobility of the transistor is the sum of the mobilities of the different conduction surfaces controlled by the gate, weighted by the respective dimensions of these surfaces. This relationship can be easily demonstrated by considering that conduction is independent of each other and that the gates are all equivalent (for only one gate voltage). We thus identify a general expression of mobility in a multigate transistor. The total current leaving the transistor is the sum currents of conduction surfaces:

$$I_{transistor} = \Sigma I_{surface} \tag{4.32}$$

The current of the different conduction surfaces according to the geometric ratio of these surfaces can be expressed using the classical equation of the drain current in a linear regime:

$$I_{surface} = \frac{l_s}{L} \cdot \mu_s C_{ox} \left(V_G - V_{th} - \frac{V_{ds}}{2} \right) \cdot V_{ds} \tag{4.33}$$

with l_s the width of the conduction surface, and μ_s its mobility.

$$I_{transistor} = \frac{W_{eff}}{L} \cdot \mu_{tot} C_{ox} \cdot \left(V_G - V_{th} - \frac{V_{ds}}{2} \right) \cdot V_{ds} \tag{4.34}$$

with W_{eff} the sum of the widths of the conduction surfaces, and μ_{tot} the total mobility of the transistor. The terms of Equation (4.32) are then replaced by the different expressions of current obtained. We obtain a general expression of the mobility in a transistor to multiple gates:

$$\mu_{tot} = \frac{1}{W_{eff}} \cdot \Sigma l_s \cdot \mu_s \tag{4.35}$$

In the particular case of rectangular trigate transistors with a section of dimensions W_{top} and H, we obtain the expression of the following mobility:

$$\mu_{TG} = \frac{W_{top}\mu_{top} + 2H\mu_{side}}{W_{top} + 2H} \tag{4.36}$$

with μ_{top} the mobility of the upper surface and μ_{side} the mobility of the lateral surfaces (which is assumed identical for both side surfaces). For a trigate transistor oriented in the crystalline direction <110> we can express the mobility using Equation (4.36), replacing the mobility of the surface upper μ_{top} by $\mu_{(100)}$ and the mobility of lateral surfaces μ_{side} by $\mu_{(110)}$ following the crystallographic orientation of the <110> channel, namely:

$$\mu_{TG} = \frac{W_{top}\mu_{(100)} + 2H\mu_{(110)}}{W_{top} + 2H} \tag{4.37}$$

However, for electrons, the mobility for a conduction surface (100) is better than that for a surface (110) [110] as W_{top} decreases, the ratio of the areas of conduction between the top surface (100) and the side surfaces (110) changes.

4.3.2 Effects of Channel Orientation

Measurements as a function of channel orientation for p-MOSFETs show the effect of changing the crystallographic orientation lateral surfaces of the FinFET transistors. Mobility for a transistor in <110> direction increases as W decreases as predicted by Equation (4.36) and that of transistors <100> decreases with W_{top} as for n-MOSFET, although to a lesser extent proportion, in line with a stronger contribution of the roughness of fins. It has been observed that the critical value W_{crit} of 20 nm below which the mobility decreases independently of the orientation of the channel is potentially attributed to the influence of corner effects for small widths. Low-temperature measurements confirm that the mobility limited by the roughness of the surface is strongly impacted by the orientation of the channel for devices. It has been shown that the μ_{SR} contribution for planar transistors is the same for both orientations; however, a difference between <110> and <100> orientation for p-MOSFET transistors has been reported. At 20 K, multiple low field mobility spikes for the <100> and <110> devices are observed. These peaks are the characteristic fingerprint of oscillations due to the density of 1D states [41–43].

The mobility model presented above has several limits. The limits of the model will be discussed in the following. The basic equation of the model is recalled for clarity:

$$\mu_{TG} = \frac{W_{top}\mu_{top} + 2H\mu_{side}}{W_{top} + 2H} \tag{4.38}$$

with μ_{top}, the mobility of the upper surface, and μ_{side} the mobility of the lateral surfaces, assumed identical for both side surfaces. The limit values of this equation allow us to identify two particular cases:

$$W_{top} \gg H \Rightarrow \mu_{TG} \approx \mu_{top} \tag{4.39}$$

$$H \gg W_{top} \Rightarrow \mu_{TG} \approx \mu_{side} \tag{4.40}$$

Generally, the measured mobility for a 10 µm gate width can be reasonably taken as the value of μ_{top} mobility [44]. Equation (4.38) can be used to extract the mobility of the lateral surfaces of trigate devices:

$$\mu_{side} = \mu_{eff,NW} \cdot \left(1 + \frac{W_{top}}{2H_{NW}}\right) - \frac{W_{top}}{2H_{NW}} \cdot \mu_{top} \tag{4.41}$$

We have thus extracted by this model the lateral surface mobility (μ_{side}) for all of n-MOSFET and p-MOSFET devices in the trigate configuration ($H = 10$ nm) and FinFET ($H = 24$ nm). It has been observed that the mobility of surfaces lateral (μ_{side}) thick devices in particular ($H = 24$ nm) is very close to the mobility of the conduction surface (110) measured elsewhere [44, 45], and therefore rather in agreement with the model of conduction surfaces. For n-MOSFET devices with fin height ($H = 12$ nm) nevertheless, the lateral mobility is less than the mobility (110), which indicates a model deviation for Trigate architectures with values of W and $H = 10$ nm.

The electron density profile calculated in Figure 4.7 using Schrödinger–Poisson shows that the density of a strong inversion carrier is very uneven along the side surfaces for devices with $H = 10$ nm. For this particular geometry, we see in

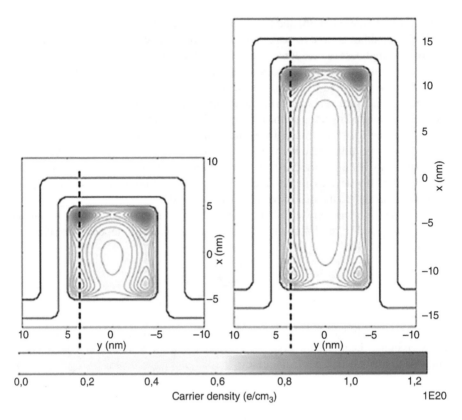

FIGURE 4.7   Poisson 2D simulation Schrödinger of the distribution of carriers within a trigate transistor for two channel heights: 10 nm and 24 nm [47].

particular that a significant part of the carriers is concentrated in the upper corners of the device for the three conduction surfaces. We can thus clearly distinguish a difference between the average carrier number and the local carrier density. On the contrary, for thicker devices (H = 24 nm), a significant part of the carrier density is constant and close to the mean value along the lateral surfaces. Consequently, for trigate, the carrier density and therefore the mobility is strongly dependent on the position in the profile of the device. This inhomogeneity due to the corner extends over about 5 nm as can be seen. The semi-analytical model of mobility is therefore no longer relevant for devices with a $W_{top} \times H$ channel width and/or height too low. 10 nm $\times$ 20 nm and would require in particular to take into account the contribution of angles [46].

4.4   SIMULATION CASE STUDIES

This section focuses on the applications of strain introduction in CMOS technology. The strain–stress relationship has been discussed using the theory of elasticity in Chapter 3. The impact of strain on the band structure of silicon is then detailed in this chapter. Piezoresistive coefficients are defined to describe the effect of stress on mobility in semiconductor materials. In the following, we consider several examples dealing with several application examples.

4.4.1 TRIGATE FINFET

The technique of inducing stress by using a tensile (for NMOS) or compressive (for p-MOSFET) SiNx capping layer is attractive because of its relatively simple process and its extendibility from bulk-Si to silicon-on-insulator MOSFETs [48]. The impact of tensile and compressive capping layers on electron and hole mobilities is investigated for Si fins with {100} sidewalls and <100> current-flow direction, and Si fins with {110} sidewalls and <110> current-flow direction, which are optimal for maximum electron and hole mobilities, respectively [49, 50]. We have generated a tensilely strained n-FinFET structure from process simulation. We have used crystal orientation (100) for the silicon wafer and the fin orientation is in the <100> direction. The FinFET with a 50 × 50 nm fin and 50 nm long channel has been virtually fabricated and shown in Figure 4.8. The fin was deposited on a SiO_2 substrate layer and a 2 nm gate isolation layer separated it from the 50 nm polysilicon gate crossing it at right angles following the CMOS process steps as shown in Figure 4.8. A 100 nm thick Si_3N_4 capping layer was deposited on top of the structure and shown in Figure 4.9. The thickness of the stress liner (Si_3N_4: CESL) over the gate is discussed in the last section as it determines the amount and the stress distribution inside the FinFET.

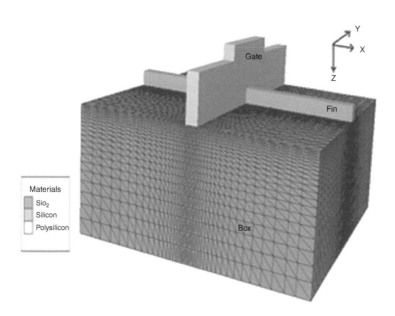

FIGURE 4.8   A FinFET with a 50 × 50 nm fin on SiO_2 buried oxide.

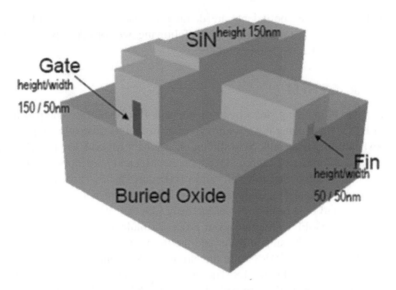

FIGURE 4.9   100 nm thick nitride cap layer deposited on top of the fin and polysilicon gate. Nominal values: BOX Thickness = 400 nm, Fin Width = 50 nm, Fin Height = 50 nm, and Gate Length = 50 nm.

After process simulation, the device structure was exported to the VictoryStress simulator [47]. The deposited nitride layer has a uniform tensile stress of +1 GPa. The stress contribution to the channel due to gate oxide has been neglected in simulation. It is also assumed that the top side of the fin has a negligible amount of current contribution. The displacement of two sidewall surfaces (left and right) and the bottom surface are set to zero. The detailed material properties used in the simulation are presented in Table 4.1. The compliance coefficient (C_{ij}) and stiffness coefficients (S_{ij}) for Si are presented in Table 4.2 has been used in this simulation. The elasticity relationships and simulation procedure (see Chapter 3) have been used in the stress simulation.

The 3D stress profiles in the FinFET induced by a tensile capping layer have been generated. A tensile CESL cap layer simultaneously introduces tensile parallel stresses and compressive vertical stress which enhances electron mobility. The stress profile for the device is shown in Figure 4.10. If we focus only on the stress in the

TABLE 4.1

Material Parameters Used in the Simulation

Materials Used	Young's Modulus (Dyne/cm²)	Poisson's Ratio	Thermal Expansion (m/m K)
silicon nitride	3.1×10^{12}	0.27	3.300×10^{6}
oxide	6.6×10^{11}	0.20	1.206×10^{7}
silicon	1.67×10^{12}	0.28	2.600×10^{16}
polysilicon	1.87×10^{12}	0.28	3.052×10^{16}

TABLE 4.2

Elasticity Compliance Coefficients C_{ij} are in GPa, and the Elastic Stiffness Coefficients S_{ij} are in 10^{-12} $m^2 \cdot N^{-1}$ Values for Si and Ge

C_{ij}	S_{ij}	Si	Ge
C_{11}		165.64	128.7
C_{12}		63.94	47.7
C_{44}		79.51	66.7
	S_{11}	0.7691	0.9718
	S_{12}	−0.2142	−0.2628
	S_{44}	1.2577	1.499

nitride layer, the stress in parallel direction varies from +140 MPa to +1.2G Pa as shown in Figure 4.10 which induces stress in the fin. The vertical direction stress profile in the nitride layer has been shown in Figure 4.11 which is found to vary from +70 MPa to +420 MPa.

For the quantitative determination of stress in the nitride layer, a 1D cutline parallel to the y-axis within the nitride layer is required. 1D cutline plot shows a uniform parallel tensile stress profile having a magnitude of 500 MPa and vertical stress of −300 MPa. This stress in the nitride layer will induce stress in the fin. Besides the nitride layer, the stress from the polysilicon also contributes to fin stress. To evaluate the stress profile in the fin we have to analyze the stress profile in the polysilicon

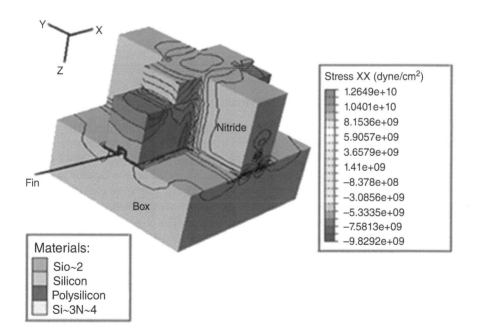

FIGURE 4.10 3D stress profile in the n-type FinFET structure used for stress simulations.

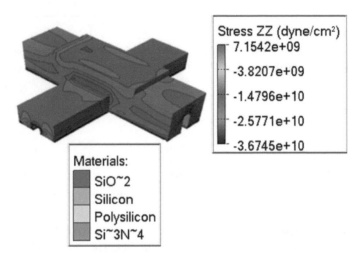

FIGURE 4.11   Vertical direction of the stress field in the capping nitride layer.

layer. The 1D stress profile in the polysilicon layer shows that vertical stress is compressive and significantly high as −2.2 GPa whereas the stress in the parallel direction is insignificant. It is assumed that a thin gate oxide layer will have a negligible effect on the stress transfer from the capping layer to the channel, and so it was not included in the simulated structure, for simplicity. As our focus is to determine the stress in the channel or fin, we have calculated the stress in the fin. A 2D von Mises stress profile is shown in Figure 4.12 which shows that the parallel stress is nonuniform. 2D stress XX profile is shown in Figure 4.13 which shows that the stress is nonuniform. The contribution of the top fin surface to FinFET current is assumed to be negligible [51].

4.4.2 Enhancement of Electron Mobility

The two components, basically S_{xx} (longitudinal) tensile stress along with it the compressive vertical stress (S_{zz}) enhance the electron mobility. The enhancement in electron mobility has been calculated. Figure 4.14 shows that electron mobility can be greatly enhanced by over 26% due to stress-induced by the tensile capping layer.

The aim of introducing stress in the device is to obtain improved electrical performance. Once stress simulation is performed, all the electrodes definition is established to evaluate the device's electrical performance. The physical models discussed in Chapter 3 have been used to investigate the effect of stress on the transistors' transfer and output characteristics. Figure 4.15a shows the variation of drain current with the variation of gate voltage at $V_d = 0.1$ V. A 60% enhancement in drain current can be observed with stress compared to the no-stress condition. The enhancement in drain current with stress is due to the enhancement in mobility following the piezo-resistance mobility relations. Similarly, the output characteristics have been shown in Figure 4.15b at $V_g = 1$ V and 2 V for both the "no stress" and "stressed condition" of

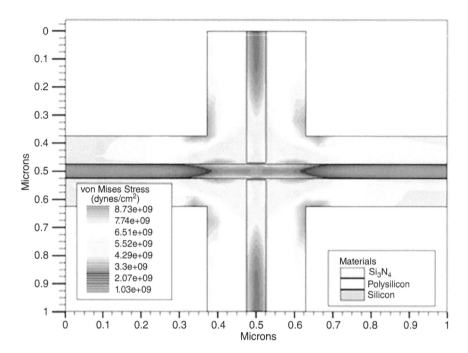

FIGURE 4.12 2D von Mises stress profile in the fin.

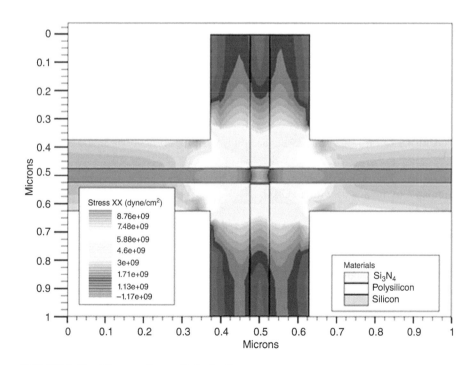

FIGURE 4.13 2D stress S_{xx} profile in the fin.

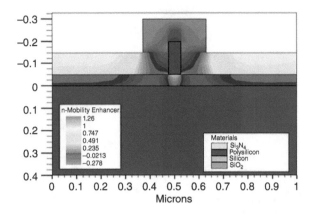

FIGURE 4.14   Electron mobility enhancement factor in the fin (contour plot).

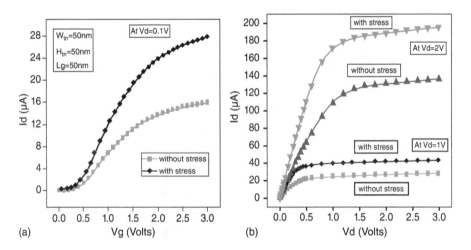

FIGURE 4.15   (a) I_d-V_g characteristics and (b) I_d-V_d characteristics of the n-FinFETs.

the device. The enhancement in I_{Dsat} is observed to be 55% in devices with stress compared to the device with no CESL.

4.4.3   FinFET Inverter

In planar CMOS technologies, orientation-dependent mobility enhancement has been demonstrated through the use of hybrid orientation technology (HOT) [52, 53]. Hole mobility increases significantly when the channel orientation changed from Si (100) to Si (110). However, the electron mobility is severely degraded with the same orientation change. However, a correlation of long channel mobility to short channel performance is necessary for a clear understanding of the performance characteristics. Moreover, the implications of sidewall surface orientation on reliability issues such as hot carriers [54] and bias temperature instability [55] also need to be addressed.

FinFETs can be fabricated with either the (110) sidewall surface or (100) sidewall where the crystal orientation of the fin sidewalls can have an impact on mobility and thereby provide a mobility boost based on orientation [28, 56]. Also, the surface orientation or fin structure may be more susceptible to degradation during stressing. New insights about the carrier transport on all the three surfaces in trigate FinFETs are necessary to obtain a clear understanding of the FinFET operation [57–59]. However, it has been shown that the model treating the trigate FinFET in terms of the (100) top and (110) lateral channels is not accurate for describing the transport properties in real FinFETs with relatively narrow fins and at low and moderate inversion charge densities [60]. This is due to different inversion carrier distributions (nonidentical scattering rates for various fin widths). Also, at high charge densities, due to fin rounding which results in ambiguous crystallographic orientation, different from the postulated (100) and (110), description of the transport properties becomes complicated [40, 61].

The strain has been implemented intentionally as a mobility booster in devices that went into production in 2004 [62]. Besides the salient effects on the charge carrier mobilities, stress affects the width of the bandgap and has widespread direct and indirect effects on the defects in semiconductors. As the device width shrinks, 3D simulation has been routinely employed in mechanical stress investigations [63]. The most common method of introducing desirable stresses into a transistor channel region is the deposition of high tensile or high compressive films of nitride-type materials [64]. This is called global stress engineering. In local stress engineering, regions could be created, for example, by epitaxial growth of compositional SiGe or SiC materials [65–69].

It is important to know how stresses from different sources interact with each other in areas of interest, for example, under the gates of various devices in an inverter cell. The interaction effects could be even more pronounced when the critical dimensions of individual devices and distances between them scale down. It is therefore the aim of this simulation case study is to investigate the 3D stress by process/device simulation. Stress has significant influences on electrical performance and hence where it should be optimized. To investigate stress effects, we used VictoryStress simulations in a FinFET inverter test structure. The inverter cell consists of three FinFETs: one FinFET is of n-type and the other two are p-type devices located parallel to each other. Both thin film liners (compressive for p-type devices, tensile for n-type devices) and stressed S/D regions (compressive for p-type devices, tensile for n-type devices) were implemented near the gate which significantly increases corresponding carrier mobility of all types of devices. The piezoresistivity mobility model discussed in Chapter 3 has been widely discussed in the literature [70–72] and is used in the simulation.

To investigate the mobility enhancement over a wider stress range, it is necessary to model the mobility enhancement in embedded SiGe source/drain p-MOSFETs under high-stress conditions. The shape of the embedded $Si_{0.8}Ge_{0.2}$ source/drain has been found to exert compressive stress in the channel ranging from 200 MPa to 1.5 GPa [73]. The channel stress is calculated using a 2D process simulation where all intentional and unintentional stress sources and stress evolution during the entire process flow are taken into account. While the channel stress induced by the compressive e-SiGe is primarily longitudinal along the channel (σ_{xx}), a much smaller, but non-negligible, compressive transverse stress across the channel (σ_{zz}) also is present. Since the effects of the longitudinal and the transverse stress components on the hole mobility are comparable in magnitude, both of these in-plane components must be considered when estimating the stress enhanced mobility. The measurements indicate the mobility enhancement increases approximately linearly at low effective compressive channel stress but becomes slightly linear above 1 GPa. The best-fit piezoresistance coefficient over the entire stress range is 93×10^{-11} Pa^{-1}, which is 30% larger than the longitudinal bulk piezoresistance value. Typical hole mobility enhancement of 140% corresponds to a compressive longitudinal stress of 1.45 GPa and compressive transverse stress of 270 MPa [74].

It has been shown that a simple piezoresistance mobility model can describe the stress impact on transistor performance with good accuracy. The piezoresistance model provides accurate stress-dependent mobility values (within about 20%) at stress levels below 1 GPa. At higher stress levels, holes exhibit linear mobility gain with increasing stress; whereas, electron mobility gain with stress becomes sublinear and eventually saturates. The following equations are used for the calculation of n- and p-mobility enhancement factors for (100)/<100> and (110)/<100> crystallographic orientations in silicon [75]:

$$\mu_n 100 = 1.0 - \left(-1.200 * \sigma_{xx} + 0.534 * \sigma_{yy} + 0.534 * \sigma_{zz}\right)$$
$$\mu_p 100 = 1.0 - \left(0.066 * \sigma_{xx} - 0.011 * \sigma_{yy} - 0.011 * \sigma_{zz}\right)$$
$$\mu_n 110 = 1.0 - \left(-0.311 * \sigma_{xx} - 0.175 * \sigma_{yy} + 0.534 * \sigma_{zz}\right)$$
$$\mu_p 110 = 1.0 - \left(0.718 * \sigma_{xx} - 0.663 * \sigma_{yy} - 0.011 * \sigma_{zz}\right)$$

In these equations, the stresses are in units of GPa, and coefficients of piezoresistivity are in 1/GPa. Considering that the piezoresistance model is a linear superposition of the contributions from the principal stress components, one can focus on optimizing the transistor to achieve the optimal stress pattern in the channel. However, the most beneficial stress components are different for n- and p-MOSFETs. The n-MOSFET benefits most from compressive vertical stress and also somewhat from tensile stress in the lateral directions; whereas, the p-MOSFET benefits most from the compressive stress in the direction along the channel. It is important to simulate stresses not just in an individual device but in the whole cell. In the following, a combination of 3D process simulator VictoryCell and 3D stress simulator VictoryStress will be employed for stress analysis in an inverter cell structure.

In general, stress simulations are performed in separate devices ignoring the proximity effects. It is important to simulate stresses not just in an individual device structure but in the whole cell. To account for proximity effects, simulation of the whole cell is necessary for which both the global and local stress engineering are considered for each device type. It is important to study stress effects on carrier mobilities of individual n-FinFET and p-FinFET devices as well as on characteristics of an inverter cell consisting of one n-FinFET and two p-FinFETs.

The schematic of the FinFET cell inverter is shown in Figure 4.16 with its top and side views. It consists of three devices named n-FinFET, p1-FinFET, and p2-FinFET

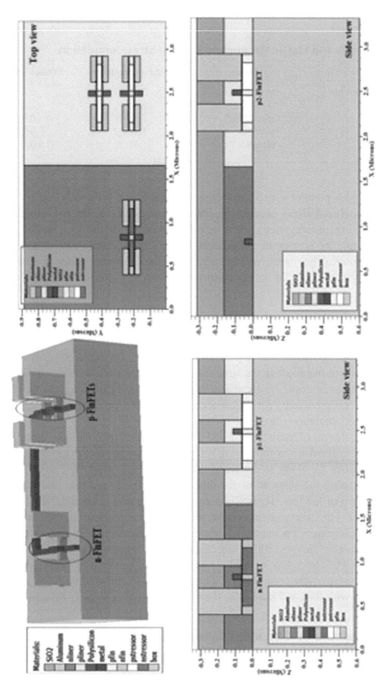

FIGURE 4.16   3D schematic view of FinFET cell inverter with its cross-section views.

TABLE 4.3

Materials and Elastic Parameters Used in Stress Simulation

Material	Name	Young's Modulus	Poisson's Ratio
Silicon nitride	"nliner"/"pliner"	3.89e12	0.3
oxide	"box"	6.6e12	0.2
silicon	"nfin"/"pfin"	1.67e12	0.28
silicon	"nstressor"	1.67e12	0.28
silicon	"pstressor"	1.67e12	0.28

where the p1 is parallel with p2 and aligned with n-FinFET. For this FinFET inverter, we have analyzed stress because of different issues like in the n-FinFET such as manufacturing process steps, corresponding material depositions with a specific thickness, and the temperature condition. Hence, the effect of stress depends on the device's material, geometry, and temperatures. The 2D stress profile is observed for all three types of FinFETs in the cell inverter.

4.4.4 STRESS SIMULATION

By default the absolute value of the initial intrinsic stress is assumed to be 1 GPa; for the n-FETs it is positive; for the p-FETs it is negative. The material parameters used for user-defined materials in the structure are shown in Table 4.3. VictoryStress [75] considers the anisotropic elastic properties of silicon as the quantities in the elasticity tensor vary insignificantly for different crystallographic orientations.

Since the simulation results are in 3D, the only convenient way of analyzing them is to extract figures of merit from inside using cut planes within the 3D structure. The five planes selected are as follows: first four planes are parallel to the fin length; two of those planes are along sidewalls of the 1st n-FinFET and 1st p-FinFET; next two planes are along the sidewall of the other 2nd p-FinFET, and the final (the fifth) is along the top of the fins. These 2D planes are generated using the "-cut" option of Tonyplot3D. The exact position of the cut was set manually using the Cutplane setting in the Cutplane view menu of Tonyplot3D [76]. Average stress calculations were performed using "factor" (100 in this case) which is required for obtaining integrated stresses in GPa. The S_{xx}-, S_{yy}-, and S_{zz}-components in the sidewalls and top side of the fin under the gate were computed by the integration of corresponding stress distributions along with the cut 5 nm below the fin-oxide boundary under the gate.

4.4.5 ORIENTATION-DEPENDENT MOBILITY EXTRACTION

Although the mechanisms behind the enhancement in stress/strain-induced mobility are fairly well understood qualitatively, quantitative evaluation is much more difficult as the type of mechanical stress-induced is indeed very complex. Various physical and numerical models for stress/strain as discussed in Chapter 3 (Section 3.1) were used for stress/strain simulations. In the simulation, mobility enhancement models based on piezoresistive coefficients discussed in this chapter (Section 4.2) were used.

TABLE 4.4
Orientation-Dependent Piezoresistivity Coefficients Used for Mobility Enhancement Calculations

n-FinFET

Orientation	x	y	z	unit
(100)/<100>	−1.022	0.534	0.534	cm²/GPa
(110)/<100>	−0.311	−0.175	0.534	cm²/GPa
p-FinFET				
Orientation	x	y	z	unit
(100)/<100>	0.066	−0.011	−0.011	cm²/GPa
(110)/<100>	0.718	−0.663	−0.011	cm²/GPa

The mobility enhancement depends on silicon crystal orientation because the tensor of piezoelectricity varies significantly for different crystallographic orientations. Orientation-dependent piezoresistivity coefficients used for mobility enhancement calculations are shown in Table 4.4. Mobility enhancement factors were calculated at the sidewall and top of the fin under the gate. Mobility enhancements for n-finFET with 2 fin orientations were found to be dependent on fin orientation, and also the enhancement factors are different because the piezoresistivity coefficients are different. For the simulation of a p-finFET device having the same structure as n-finFET, the stresses can be calculated using the same procedure but with different piezoresistivity coefficients.

The stress distribution profiles show positive tensile stress and negative compressive stress in the n- and p-FinFET, respectively. The induced stress profile is nonuniform. The average stress calculated for individual FinFETs in the cell has been summarized in Table 4.5. It is seen that the stress components differ in stresses and enhancements obtained from the simulation of individual devices (n-FinFET). This happens due to the interaction between n-liner and p-liner. For a non-planar transistor structure such as the FinFET, main stress components, S_{xx} (along the channel), has a significant effect on carrier mobility. The other two components S_{yy} (vertical to the channel surface) and S_{zz} (across the channel width) can also be significant and hence their effects on mobility must be taken into account.

TABLE 4.5
Calculated Mobility Enhancement Factors for n- and p-FinFETs

First p-FinFET	(100)/<100>	(110)/<100>
side1_p1	1.00215	0.910603
Side2_p1	1.00137	0.899863
top_p1	1.00181	0.905628
Second p-FinFET	(110)/<100>	(100)/<100>
side1_p1_2	0.910603	1.00195
side2_p1_2	0.899863	1.00289
top_p1_2	0.905628	1.00248

TABLE 4.6

Average Stress Calculation of Inverter for Individual FinFETs

Device	S_{xx}	S_{yy}	µn_100	µn_110	µp1_100	µp_110
n-FinFET	0.058	−0.225	1.177	0.979		
p1-FinFET	−0.009	0.205	-	-	1.002	1.137
p2-FinFET	−0.022	0.216	-	-	1.004	1.166

The electron mobility in a (100)-sidewall Fin/trigate FET can be greatly enhanced by a tensile capping layer, due largely to the significant induced compressive S_{yy} and S_{zz} as well as the tensile S_{xx}. Although hole mobility in a (110) sidewall fin is degraded due to the tensile S_{xx} induced by a compressive capping layer, it is enhanced more so by the large induced tensile S_{yy} so that the net impact of a 1 GPa compressive capping layer on a (110)-sidewall fin is to enhance the hole mobility by a modest amount (less than 25%). The enhancement in mobility due to each side of the fin in the inverter cell has been investigated and summarized in Table 4.6. A tensile capping layer is expected to provide enhancement (>100%) in electron mobility for a (100)-sidewall fin, while a compressive capping layer is expected to provide a modest amount (<25%) of hole mobility enhancement for a (110)-sidewall fin.

4.4.6 SUMMARY

In this chapter, we have presented the possible performance enhancement achievable via technology CAD simulations and demonstration of CMOS FinFETs. We considered in particular, the evolution of the piezoresistive effect and electrical transport properties in field-effect transistors involving different variables such as the geometry, temperature, and internal mechanical stress, to study the effects of the reduction at the extreme scaling limit of the channel and gate dimensions in the transistors. We have performed stress analyses in Si nanoscale FinFETs. We have performed the mobility extraction of trigate FinFETs using piezoresistive models [77] compared to planar devices in agreement with better mobility on the fins despite a greater roughness due to the lithography process. The proposed semi-analytical model of mobility offers the possibility of interpreting and modeling the variations of the mobility with the geometry of the channel, as well as the number of the variations of gates. This model allowed us to highlight the effect of the ultrasmall dimensions on conduction in multigate transistors. These reduced dimensions lead to early confinement due to additional gates and abrupt geometry between them. These two factors create a strong inhomogeneity of the density of carriers on the conduction surfaces. This inhomogeneity is generally not taken into account in the standard description of the current and deviates from experimental measurements, especially for sub-10 nm devices.

In this simulation case study, we used the 3D simulator VictoryStress to analyze stress effects on carrier mobilities of individual n- and p-FinFET devices. We demonstrated that a combination of the 3D process simulator VictoryCell and 3D stress simulator VictoryStress allow fast and accurate stress analysis of complex cell structures, such as inverter cells. We have performed detailed analysis and optimization of various stress engineering schemes by varying geometrical characteristics, orientation, and material composition of each device as well as by changing the location and density of the individual devices inside the cell layout.

REFERENCES

[1] J. M. Luttinger, "Quantum theory of cyclotron resonance in semiconductors: General theory," *Phys. Rev.*, vol. 102, no. 4, pp. 1030–1041, 1956, doi: 10.1103/PhysRev.102.1030.

[2] S. E. Thompson et al., "A 90-nm logic technology featuring strained-silicon," *IEEE Trans. Electron Dev.*, vol. 51, no. 11, pp. 1790–1797, 2004, doi: 10.1109/TED.2004.836648.

[3] T. Skotnicki, J. A. Hutchby, T. J. King, H. S. P. Wong, and F. Boeuf, "The end of CMOS scaling: Toward the introduction of new materials and structural changes to improve MOSFET performance," *IEEE Circuits Devices Mag.*, vol. 21, no. 1, pp. 16–26, 2005, doi: 10.1109/MCD.2005.1388765.

[4] T. Skotnicki, "MASTAR manual, ver. 4." 2005.

[5] G. L. Bir and G. E. Pikus, *Symmetry and Strain-Induced Effects in Semiconductors.* Wiley, New York, 1974.

[6] J. M. Luttinger and W. Kohn, "Motion of electrons and holes in perturbed periodic fields," *Phys. Rev.*, vol. 97, no. 4, pp. 869–883, 1955, doi: 10.1103/PhysRev.97.869.

[7] Y. Sun, S. E. Thompson, and T. Nishida, "Physics of strain effects in semiconductors and metal-oxide-semiconductor field-effect transistors," *J. Appl. Phys.*, vol. 101, no. 10, 2007, doi: 10.1063/1.2730561.

[8] G. Bidal et al., *"First CMOS integration of Ultra Thin Body and BOX (UTB2) structures on bulk Direct Silicon Bonded (DSB) wafer with multi-surface orientations,"* in *Technical Digest - International Electron Devices Meeting, IEDM*, 2009, pp. 677–680, doi: 10.1109/IEDM.2009.5424247.

[9] K. Uchida, A. Kinoshita, and M. Saitoh, *"Carrier transport in (110) nMOSFETs: Subband structures, non-parabolicity, mobility characteristics, and uniaxial stress engineering,"* in *2006 International Electron Devices Meeting*, 2006, pp. 1–3, doi: 10.1109/IEDM.2006.346943.

[10] K. Shin et al., "Study of bending-induced strain effects on MuGFET performance," *IEEE Electron Device Lett.*, vol. 27, no. 8, pp. 671–673, 2006, doi: 10.1109/LED.2006.878047.

[11] T. Irisawa, T. Numata, T. Tezuka, N. Sugiyama, and S. I. Takagi, *"Electron transport properties of ultrathin-body and tri-gate SOI nMOSFETs with biaxial and uniaxial strain,"* in *Technical Digest - International Electron Devices Meeting, IEDM*, 2006, pp. 1–4, doi: 10.1109/IEDM.2006.346811.

[12] K. Uchida and M. Saitoh, *"Stress engineering in (100) and (110) nMOSFETs,"* in *2008 9th International Conference on Solid-State and Integrated-Circuit Technology*, 2008, pp. 109–112, doi: 10.1109/ICSICT.2008.4734485.

[13] V. Destefanis et al., "Structural properties of tensile strained Si layers grown on SiGe(100), (110), and (111) virtual substrates," *J. Appl. Phys.*, vol. 106, no. 4, p. 43508, 2009, doi: 10.1063/1.3187925.

[14] S. Bangsaruntip et al., *"Gate-all-around silicon nanowire 25-stage CMOS ring oscillators with diameter down to 3 nm,"* in *Digest of Technical Papers - Symposium on VLSI Technology*, 2010, pp. 21–22, doi: 10.1109/VLSIT.2010.5556136.

[15] K. Tachi et al., *"Experimental study on carrier transport limiting phenomena in 10 nm width nanowire CMOS transistors,"* in *2010 International Electron Devices Meeting*, 2010, pp. 34.4.1–34.4.4, doi: 10.1109/IEDM.2010.5703476.

[16] S. Narasimha et al., "22nm High-performance SOI technology featuring dual-embedded stressors, Epi-Plate High-K deep-trench embedded DRAM and self-aligned Via 15LM BEOL," 2012, doi: 10.1109/IEDM.2012.6478971.

[17] J. M. Hartmann et al., "Mushroom-free selective epitaxial growth of Si, SiGe, and SiGe: B raised sources and drains," *Solid-State Electronics*, 2013, vol. 83, pp. 10–17, doi: 10.1016/j.sse.2013.01.033.

[18] R. Coquand, "Démonstration de l'intérêt des dispositifs multi-grilles auto-alignées pour les nœuds sub-10nm." PhD Thesis, Univ. Grenoble Alpes, 2013.

[19] O. Weber et al., *"Examination of additive mobility enhancements for uniaxial stress combined with biaxially strained Si, biaxially strained SiGe and Ge channel MOSFETs,"* in *2007 IEEE International Electron Devices Meeting*, 2007, pp. 719–722, doi: 10.1109/IEDM.2007.4419047.

[20] F. Rochette et al., "Piezoresistance effect of strained and unstrained fully-depleted silicon-on-insulator MOSFETs integrating a HfO₂/TiN gate stack," *Solid. State. Electron.*, vol. 53, no. 3, pp. 392–396, 2009, doi: 10.1016/j.sse.2009.01.017.

[21] C. S. Smith, "Piezoresistance effect in germanium and silicon," *Phys. Rev.*, vol. 94, no. 1, pp. 42–49, 1954, doi: 10.1103/PhysRev.94.42.

[22] J. F. Nye, *Physical Properties of Crystals - Their Representation by Tensors and Matrices*. Oxford Science Publications, 1985.

[23] M. Casse et al., *"Study of piezoresistive properties of advanced CMOS transistors: Thin film SOI, SiGe/SOI, unstrained and strained Tri-Gate Nanowires,"* in *2012 International Electron Devices Meeting*, 2012, pp. 28.1.1–28.1.4, doi: 10.1109/IEDM.2012.6479119.

[24] Y. Kanda, "A graphical representation of the piezoresistance coefficients in silicon," *IEEE Trans. Electron Devices*, vol. 29, no. 1, pp. 64–70, 1982, doi: 10.1109/T-ED.1982.20659.

[25] F. Rochette, "Study and characterization of the mechanical stress influence on electronic transport properties in advanced MOS architectures." PhD Thesis, Institut National Polytechnique de Grenoble, 2008.

[26] K. Uchida et al., *"Physical mechanisms of electron mobility enhancement in uniaxial stressed MOSFETs and impact of uniaxial stress engineering in ballistic regime,"* in *IEEE InternationalElectron Devices Meeting, 2005. IEDM Technical Digest.*, 2005, pp. 129–132, doi: 10.1109/IEDM.2005.1609286.

[27] B. DeSalvo et al., *"A mobility enhancement strategy for sub-14nm power-efficient FDSOI technologies,"* in *2014 IEEE International Electron Devices Meeting*, 2014, pp. 7.2.1–7.2.4, doi: 10.1109/IEDM.2014.7047002.

[28] J. Pelloux-Prayer et al., *"Transport in TriGate nanowire FET: Cross-section effect at the nanometer scale,"* in *2016 IEEE SOI-3D-Subthreshold Microelectronics Technology Unified Conference (S3S)*, 2016, pp. 1–2, doi: 10.1109/S3S.2016.7804374.

[29] J. Pelloux-Prayer et al., *"Strain effect on mobility in nanowire MOSFETs down to 10nm width: Geometrical effects and piezoresistive model,"* in *2015 45th European Solid State Device Research Conference (ESSDERC)*, 2015, vol. 2015-Novem, pp. 210–213, doi: 10.1109/ESSDERC.2015.7324752.

[30] M. Lundstrom, *Fundamentals of Carrier Transport*. Addison-Wesley Publishing Company, Reading, MA, 1990.

[31] M. Cassé et al., "Carrier transport in HfO₂/metal gate MOSFETs: Physical insight into critical parameters," *IEEE Trans. Electron Devices*, vol. 53, no. 4, pp. 759–768, 2006, doi: 10.1109/TED.2006.870888.

[32] M. V. Fischetti et al., "Band structure, deformation potentials, and carrier mobility in strained Si, Ge, and SiGe alloys," *J. Appl. Phys.*, vol. 80, no. 4, pp. 2234–2252, 1996, doi: 10.1063/1.363052.

[33] A. Matthiessen and C. Vogt, "IV. On the influence of temperature on the electric conducting-power of alloys," *Philos. Trans. R. Soc. London*, vol. 154, pp. 167–200, 1864, doi: 10.1098/rstl.1864.0004.

[34] S. Takagi et al., "On the universality of the inversion layer mobility in Si MOSFETs: Part I: Effects of Surface Orientation," *IEEE Trans. Electron Dev.*, vol. 41, pp. 2363–2368, 1994, doi: 10.1109/16.337450.

[35] S. I. Takagi, J. L. Hoyt, J. J. Welser, and J. F. Gibbons, "Comparative study of phonon-limited mobility of two-dimensional electrons in strained and unstrained Si metal-oxide-semiconductor field-effect transistors," *J. Appl. Phys.*, vol. 80, no. 3, pp. 1567–1577, 1996, doi: 10.1063/1.362953.

[36] M. O. Baykan, S. E. Thompson, and T. Nishida, "Strain effects on three-dimensional, two-dimensional, and one-dimensional silicon logic devices: Predicting the future of strained silicon," *J. Appl. Phys.*, vol. 108, no. 9, 2010, doi: 10.1063/1.3488635.

[37] T. Ando, A. B. Fowler, and F. Stern, "Electronic properties of two-dimensional systems," *Rev. Mod. Phys.*, vol. 54, no. 2, pp. 437–672, 1982, doi: 10.1103/RevModPhys.54.437.

[38] L. Thevenod et al., "Influence of TiN metal gate on Si/SiO_2 surface roughness in N and PMOSFETs," *Microelectronic Engineering*, 2005, vol. 80, no. SUPPL., pp. 11–14, doi: 10.1016/j.mee.2005.04.037.

[39] J. P. Colinge, "The SOI MOSFET: From single gate to multigate," *FinFETs and Other Multi-Gate Transistors*. Springer US, pp. 1–48, 2008, doi: 10.1007/978-0-387-71752-4_1.

[40] T. Rudenko, V. Kilchytska, N. Collaert, M. Jurczak, A. Nazarov, and D. Flandre, "Carrier mobility in undoped triple-gate FinFET structures and limitations of its description in terms of top and sidewall channel mobilities," *IEEE Trans. Electron Devices*, vol. 55, no. 12, pp. 3532–3541, 2008, doi: 10.1109/TED.2008.2006776.

[41] J. P. Colinge et al., "Low-temperature electron mobility in trigate SOI MOSFETs," *IEEE Electron Device Lett.*, vol. 27, no. 2, pp. 120–122, 2006, doi: 10.1109/LED.2005.862691.

[42] R. Coquand et al., "*Low-temperature transport characteristics in SOI and sSOI nanowires down to 8nm width: Evidence of IDS and mobility oscillations*," in *European Solid-State Device Research Conference*, 2013, pp. 198–201, doi: 10.1109/ESSDERC.2013.6818853.

[43] K. S. Yi, K. Trivedi, H. C. Floresca, H. Yuk, W. Hu, and M. J. Kim, "Room-temperature quantum confinement effects in transport properties of ultrathin si nanowire field-effect transistors," *Nano Lett.*, vol. 11, no. 12, pp. 5465–5470, 2011, doi: 10.1021/nl203238e.

[44] M. Koyama et al., "*Study of carrier transport in strained and unstrained SOI tri-gate and omega-gate Si-nanowire MOSFETs*," in *2012 Proceedings of the European Solid-State Device Research Conference (ESSDERC)*, 2012, pp. 73–76, doi: 10.1109/ESSDERC.2012.6343336.

[45] T. Signamarcheix et al., "Fully depleted silicon on insulator MOSFETs on (1 1 0) surface for hybrid orientation technologies," *Solid-State Electronics*, vol. 59, no. 1, pp. 8–12, 2011, doi: 10.1016/j.sse.2011.01.013.

[46] Z. Zeng, F. Triozon, S. Barraud, and Y. Niquet, "A simple interpolation model for the carrier mobility in trigate and gate-all-around silicon NWFETs," *IEEE Trans. Electron Devices*, vol. 64, no. 6, pp. 2485–2491, 2017, doi: 10.1109/TED.2017.2691406.

[47] J. Pelloux-Prayer, "Etude expérimentale des effets mécaniques et géométriques sur le transport dans les transistors nanofils à effet de champ." PhD Thesis, Université Grenoble Alpes, 2017.

[48] C. D. Sheraw et al., "*Dual stress liner enhancement in hybrid orientation technology*," in *IEEE VLSI Tech. Symp. Dig.*, 2005, pp. 12–13, doi: 10.1109/.2005.1469192.

[49] H. Irie et al., "In-plane mobility anisotropy and universality under uni-axial strains in nand p-MOS inversion layers on (100), [110], and (111) Si," *IEEE Int. Electron Devices Meet.*, pp. 225–228, 2004, doi: 10.1109/IEDM.2004.1419115.

[50] T. Satô, Y. Takeishi, H. Hara, and Y. Okamoto, "Mobility anisotropy of electrons in inversion layers on oxidized silicon surfaces," *Phys. Rev. B*, vol. 4, no. 6, pp. 1950–1960, 1971, doi: 10.1103/PhysRevB.4.1950.

[51] J. G. Fossum et al., "Pragmatic design of nanoscale multi-gate CMOS," *IEEE IEDM Tech. Dig.*, 2004, pp. 613–616, doi: 10.1109/IEDM.2004.1419236.

[52] M. Yang et al., "Hybrid-Orientation Technology (HOT): Opportunities and challenges," *IEEE Trans. Electron Dev.*, vol. 53, pp. 965–978, 2006, doi: 10.1109/TED.2006.872693.

[53] K. Shin, C. O. Chui, and T. J. King, *"Dual stress capping layer enhancement study for hybrid orientation FinFET CMOS technology,"* in *Technical Digest - International Electron Devices Meeting, IEDM*, 2005, vol. 2005, pp. 988–991, doi: 10.1109/iedm.2005.1609528.

[54] Y. K. Choi, D. Ha, E. Snow, J. Bokor, and T. J. King, *"Reliability study of CMOS FinFETs,"* in *Technical Digest - International Electron Devices Meeting*, 2003, pp. 177–180, doi: 10.1109/iedm.2003.1269206.

[55] C. D. Young et al., *"Improved interface characterization technique for high-k/metal gated MugFETs utilizing a gated diode structure,"* in *Proceedings of 2010 International Symposium on VLSI Technology, System and Application, VLSI-TSA 2010*, 2010, pp. 68–69, doi: 10.1109/VTSA.2010.5488943.

[56] M. Casse et al., *"An improved mobility model for FDSOI TriGate and other multi-gate Nanowire MOSFETs down to nanometer-scaled dimensions,"* in *2017 IEEE SOI-3D-Subthreshold Microelectronics Technology Unified Conference (S3S)*, 2017, vol. 2018, pp. 1–3, doi: 10.1109/S3S.2017.8309241.

[57] V. V. Iyengar, A. Kottantharayil, F. M. Tranjan, M. Jurczak, and K. De Meyer, "Extraction of the top and sidewall mobility in FinFETs and the impact of fin-patterning processes and gate dielectrics on mobility," *IEEE Trans. Electron Devices*, vol. 54, no. 5, pp. 1177–1184, 2007, doi: 10.1109/TED.2007.894937.

[58] A. V. Thathachary et al., *"Impact of sidewall passivation and channel composition on InxGa1-xAs FinFET performance,"* *IEEE Electron Device Lett.*, vol. 36, no. 2, pp. 117–119, 2015, doi: 10.1109/LED.2014.2384280.

[59] C. D. Young et al., "Performance and reliability investigation of (110) and (100) sidewall oriented MugFETs," 2012, p. 1, doi: 10.1109/isdrs.2011.6135249.

[60] C. D. Young et al., *"Critical discussion on (100) and (110) orientation dependent transport: nMOS planar and FinFET,"* in *2011 Symposium on VLSI Technology - Digest of Technical Papers*, 2011, pp. 18–19.

[61] S. Takagi et al., "Device structures and carrier transport properties of advanced CMOS using high mobility channels," *Solid. State. Electron.*, vol. 51, no. 4 spec. iss., pp. 526–536, 2007, doi: 10.1016/j.sse.2007.02.017.

[62] S. E. Thompson et al., "A 90-nm logic technology featuring strained-silicon," *IEEE Trans. Electron Devices*, vol. 51, no. 11, pp. 1790–1797, Nov. 2004, doi: 10.1109/TED.2004.836648.

[63] G. A. Armstrong and C. K. Maiti, *TCAD for Si, SiGe, and GaAs Integrated Circuits*. The Institution of Engineering and Technology (IET), UK, 2008.

[64] S. Pidin et al., *"A novel strain enhanced CMOS architecture using selectively deposited high tensile and high compressive silicon nitride films,"* in *IEEE IEDM Tech. Dig.*, 2004, pp. 213–216, doi: 10.1109/iedm.2004.1419112.

[65] F. Nouri et al., *"A systematic study of trade-offs in engineering a locally strained pMOS-FET,"* in *IEEE IEDM Tech. Dig.*, 2004, pp. 1055–1058, doi: 10.1109/IEDM.2004.1419378.

[66] J. Wang et al., *"Novel channel-stress enhancement technology with eSiGe S/D and recessed channel on damascene gate process,"* in *Digest of Technical Papers - Symposium on VLSI Technology*, 2007, pp. 46–47, doi: 10.1109/VLSIT.2007.4339721.

[67] A. Madan, G. Samudra, and Y. C. Yeo, "Strain optimization in ultrathin body transistors with silicon-germanium source and drain stressors," *J. Appl. Phys.*, vol. 104, no. 8, p. 84505, Oct. 2008, doi: 10.1063/1.3000481.

[68] T. Numata et al., *"Performance enhancement of partially- and fully-depleted strained-SOI MOSFETs and characterization of strained-Si device parameters,"* in *IEEE IEDM Tech. Dig.*, 2004, pp. 177–180, doi: 10.1109/IEDM.2004.1419100.

[69] S. Flachowsky et al., *"Stress memorization technique for n-MOSFETs. Where is the stress memorized?"* in *Ultimate Integration on Silicon, ULIS2010*, Glasgow, 2010, pp. 149–152.

[70] C. Chen et al., "Analysis of Ultrahigh apparent mobility in oxide field-effect transistors," *Adv. Sci.*, vol. 6, no. 7, p. 1801189, 2019, doi: 10.1002/advs.201801189.

[71] D. Colman, J. P. Mize, and R. T. Bate, "Orientational and azimuthal dependence of piezoresistivity and mobility in silicon inversion layers," *IEEE Trans. Electron Devices*, vol. 14, no. 9, p. 631, 1967, doi: 10.1109/T-ED.1967.16048.

[72] K. Rim et al., *"Low field mobility characteristics of sub-100-nm unstrained and strained-Si MOSFETs,"* in *IEEE IEDM Tech. Dig.*, 2002, pp. 43–46, doi: 10.1109/IEDM.2002.1175775.

[73] K. Chui et al., *"Ultra-thin-body P-MOSFET featuring silicon-germanium source/drain stressors with high germanium content formed by local condensation,"* in *2006 European Solid-State Device Research Conference*, 2006, pp. 85–88, doi: 10.1109/ESSDER.2006.307644.

[74] C. K. Maiti et al., "Hole mobility enhancement in strained-Si p-MOSFETs under high vertical fields," *Solid-State Electron.*, vol. 41, no. 12, pp. 1863–1869, 1997, doi: 10.1016/S0038-1101(97)00152-4.

[75] Silvaco International, VictoryStress user manual, 2018.

[76] Silvaco International, TonyPlot3D user manual, 2018.

[77] A. Toriumi, M. Iwase, and H. Tango, "On the universality of inversion layer mobility in Si MOSFET's: Part II—effects of surface orientation," *IEEE Trans. Electron Devices*, vol. 41, no. 12, pp. 2363–2368, 1994, doi: 10.1109/16.337450.

5 Bulk-Si FinFETs

The world of microelectronics has been built on transistors. Since the first chip demonstration in 1958, now chips with billions of transistors are the most manufactured item in the world. In this chapter, we shall discuss the field-effect transistors, the switch, which is the basis of the binary logic. Its performance has been improved at every technology node to reconcile higher performance with lower power consumption. This technological feat enabled the emergence of multipurpose devices through what is now called digital convergence. To continue miniaturization and allow the production of more and more transistors on the same chip, the industry has gone through many changes to face the physical limits of their functioning. From the introduction of high permittivity (high-k) materials, the insertion of a gate metal until the use of mechanical strain to improve the performance [1]. So far, the transistors have retained their planar architecture on the substrate. Some manufacturers have attempted the integration of planar structures to the silicon-on-insulator (SOI) substrates that could reach the 10 nm technology node [2]. The 22 nm technology proposed in 2011 by Intel, however, has upset the conventions by introducing large-scale production of nonplanar and multigate transistors called the TriGate FinFET. This change of architecture represents a technological challenge for its manufacture but many advantages in its operation.

In this chapter, we shall briefly discuss the history of the transistor and the solutions proposed to overcome the limitations that miniaturization imposes [3]. We shall discuss why the technology of today requires a change in architecture. The first part of the chapter will thus be devoted to the evolution and the presentation of advanced solutions to enable the development of multigate architectures [4]. In the next section, we shall discuss the physical and electrical features of these devices, as well as the description of their operation, which will be presented. We shall show the advantages of using a vertical conduction channel in TriGate FinFET transistors compared to conventional transistors. We shall also discuss how performance can be improved through different manufacturing methods such as the introduction of mechanical stress. In Section 5.3, we shall present results from numerical simulations of the fabrication of trigate FinFET transistors on Si substrates. We shall discuss the dimensioning necessary for this type of architecture and its advantages to control the short channel effects (SCE). The manufacture of gate-all-around transistors is particularly delicate, in particular for the formation of the gate around a channel. We shall also discuss the advantages of the gate-all-around architecture on the device electrostatics.

5.1 OPERATING PRINCIPLE

The transition to nodes <90 nm with gates of <100 nm have shown the limits of conventional architecture [5] and paved the way for a technological development much more complex than the only reduction problem scale. As we have seen earlier, the characteristics of a transistor depend on physical parameters, related to materials, or related to the geometry of the transistor. The development of technological nodes has led to the definition of gate length well below 100 nm (of the order of 10 nm in 2017). On this scale, the doped zones which define the source and the drain of the transistor have a non-negligible control on the conduction channel: this is the appearance of so-called SCE. For p-n junctions formed at source–channel interfaces or drain–channel, it is seen that their space charge areas have more influence on the conduction channel. The electrical length is, therefore, less than the physical length and this difference becomes significant for sub-100 nm dimensions.

This first phenomenon reduces the potential barrier between the source and the drain, which virtually amounts to modifying channel doping or threshold voltage of the transistor (which can be defined by the equations below, see reference [6]) is all the more true that the drain voltage is large, where the threshold voltage is further reduced by the value noted for drain-induced barrier lowering (DIBL). The DIBL is thus defined by the variation of threshold voltage V_T when the drain voltage V_d is changed (normalized). We write:

$$SCE = \frac{\varepsilon_{Si}}{\varepsilon_{ox}} \frac{t_{ox}}{L_{EL}} \frac{T_{Dep}}{L_{EL}} \times \left(1 + \frac{X_j^2}{L_{EL}^2}\right) \times \frac{k_B T}{q} \ln\left(\frac{N_{ch} N_{sd}}{ni^2}\right) \quad (5.1)$$

$$DIBL = \frac{\varepsilon_{Si}}{\varepsilon_{ox}} \frac{t_{ox}}{L_{EL}} \frac{T_{Dep}}{L_{EL}} \times \left(1 + \frac{X_j^2}{L_{EL}^2}\right) \times V_d \quad (5.2)$$

$$V_{Tshort} = V_{Tlong} - SCE - DIBL \quad (5.3)$$

with ε, the relative permittivities of the materials, $t_{ox,}$ and T_{Dep} the gate oxide thicknesses and depletion, L_{EL} the electrical length, X_j the depth of the junction, N_{ch} the doping of the channel, N_{sd} the source–drain doping, and no intrinsic doping.

These effects induce a loss of control of the gate on the channel and will be all the more as the gate length becomes small. In the analytic expression Equation (5.1), the term SCE can be expressed in terms of channel doping N_{ch}. In order not to deteriorate the transport properties and to modify the threshold voltage V_T of the long transistors, the proposed solution is to use so-called pocket implantation. This process additionally forms a second doped zone and modifies the electrical length L_{EL} of the short transistors, local modification of doping, which does not affect the long-channel transistors, reduces the variation of V_T, when the length of the gate decreases. Nevertheless, the solution is not perfect and can modify the access resistances of the transistor.

With the effects of miniaturization, the drain voltage used in the circuits is decreasing toward the advantage of reduced power consumption. To keep a sufficient electric field and thus control the channel, the miniaturization also provides for a reduction in the thickness of the gate oxide. However, we have reached a physical limit, since for an oxide < 1.5 nm, the appearance of quantum effects leads to gate leakage current by the tunnel effect. The size t_{ox} must therefore be sufficiently large. However, this quantum effect could be used to improve the speeds of memories [7] or more recently in the case of TFETs [8]. To this is added the poly depletion, where the charges present in the channel influence the electric field in the gate polysilicon. It comes down virtually to move the gate away from the channel, which reduces its control over the channel. It is quantified at about 4 Å. This poly depletion zone for an n-MOSFET and the thickness oxide is virtually increased in the same proportions. On the channel side too, the presence of charge carriers at the oxide–channel interface is forbidden; however, the presence of a 4 Å zone from the physical interface causes quantification effects. This forbidden area is called dark space. To continue the race to miniaturization, new solutions have to be adopted. Their implementation upset the world of microelectronics since it no longer rests on a historical track but the use of new materials. To reduce the gate voltage while maintaining an electric field enough to control the channel, the use of SiO_2 is no longer satisfactory. The solution used from the 45 nm technology node (in mass production) brings into play a dielectric material with high permittivity, called high-k. Equivalent oxide thickness (EOT) is defined as:

$$EOT = T_{High-k} \times \frac{\varepsilon_{SiO_2}}{\varepsilon_{High-k}} \qquad (5.4)$$

with T_{High-k} the thickness of material with high permittivity, and ε their relative permittivities. With a permittivity five times greater than that of SiO_2 (layer of Hafnium-based high-k material (HfO_2 or HfSiON are predominantly used today), the five times thicker gate oxide gives rise to the same capacitance, C_{ox}. It is thus easily understood that tunneling leakage current, dependent exponentially on the thickness of materials, is no longer a major problem. But beyond the complexity of using these materials (process cost, etc.), it has been shown that the interface between the silicon channel and the high-k dielectric is of poor quality, mechanically fragile, and electrically susceptible to trapping. The use of SiO_2 is therefore still essential at the interface. This plays an important role in the quality of the transport of carriers and its thickness should be optimized.

The modification of the gate dielectric was accompanied by a stack since polysilicon is now replaced by a metallic gate. In this way, one gets rid of the poly depletion and the thickness equivalent oxide is reduced accordingly. The use of metal is not anecdotal since it must be compatible in terms of processing and the integration of high-k. The choice will also influence the output and thus on the threshold voltage of the transistor. To be compatible with the n- and p-type transistors, the choice was focused on the so-called mid-gap metals, that is to say, whose output work corresponds to equidistant energy at the gate. In this way, the characteristics of both types of transistors are made symmetrical, which improves the performance of CMOS circuits (combining both types of transistors).

5.1.1 Mechanical Stresses

The modification of the gate by the use of a high-k/metal stack changes the behavior of a transistor and carrier mobility which can be degraded by the presence of charges trapped in the dielectric [9–11]. Optimizations are still being studied for gate materials to respond to future needs dictated by the International Technology Roadmap for

Semiconductors (ITRS), such as adding materials to finely adjust the output of n-MOSFET and p-MOSFET in 32 nm technology [12]. To meet performance needs, one of the technological levers is the use of intentional mechanical stresses within a device. The strain is inherent to the manufacturing processes, due to different thermal budgets or deposition of the gate metals. The latter is particularly interesting for so-called "gate last" integrations, where the gate is built after the sources and drains. The stresses can be intentionally transmitted to the silicon of the conduction channel, to modify its band structure. It depends on the structure of the device (substrate, surface plane, and orientation).

In-depth studies of band structures show that tensile strain allows the improvement of the electron transport in the case of an n-MOSFET (conduction by the electrons). This can be related both to the alteration of the valleys of the Si but also to a reduction in the effective mass of the carriers. From the point of view of integration processes, different techniques have been proposed to control these mechanical strains, such as the thin layer deposition [13] viz., contact etch stop layer, or stress memorization [14]. In all cases, the mobility of carriers is improved and thus allows a gain on the current. Current techniques use the differences in the size of atomic masses between materials. In the case of p-MOSFET, it has been shown that compressive stress improves the transport of holes. SiGe is thus introduced into the source/drain (from the Intel 90 nm node) to compress the channel parallel to the transport and improve the mobility of the holes.

It may be noted that the effect of miniaturization makes space available around the gate reduced. This is why, to maintain a level of sufficient stress, the concentration of Ge S/D areas keeps increasing with the technology nodes [15]. In the same way, the material *SiC* has been used [16–18] to induce stress in the channel of the NMOS. Also, the use of strained SOI substrates (sSOI) provides intrinsic stress by the substrate and not by a deposited material. Historically, for reasons of cost and process compatibility, Si has been used since the beginning of microelectronics. Si is not the material having the best properties for current transport. However, the discovery of the transistor effect, and the first transistor manufactured, was in Germanium crystal [19]. As a natural extension, the SiGe source–drain is used today to improve the transport of holes the channel itself could consist of SiGe or only Ge. In the same way, alloys of more exotic materials from the family of III-V compounds could be used in the manufacturing of MOSFET transistors. Effective masses of holes (for the Ge) and electrons (for the III-V) are low enough to allow mobility well above current standards [20]. The use of these materials remains limited because of their rarity or too low bandgap that will be causing larger leakage current in the transistor.

5.1.2 CRYSTALLOGRAPHIC ORIENTATION

Silicon has a cubic crystallographic structure similar to diamond. The fabrication of silicon substrates gives access to different crystallographic planes. Besides, the rotation of substrates during manufacture also allows us to choose the direction of transport of the charge carriers. These different possibilities are shown in Figure 5.1. It will be noted that in the industry the surface (100) of the Si is used as standard and

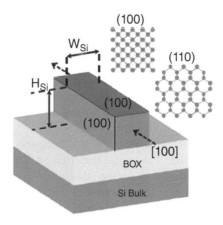

FIGURE 5.1 Schematic diagram of the active zone in silicon with the different crystalline planes [21].

the substrates are oriented to follow the <110> direction, a historic choice that combines a low cost of the substrate with good performance of NMOS. Working with the plane (111), having the highest density of Si atoms, corresponds to an alignment along a plane intersecting the diagonal of the cube of the lattice, and the plane (110) is accessible in the diagonal of a face of the lattice.

It is known that these crystallographic properties play an important role in the semiconductor properties of the material. Different studies [23–24] thus made it possible to extract the actual mobilities of the carriers according to different configurations (see Figure 5.2). It should be noted that the standard case used in the industry is optimized for an n-type transistor, but that the mobility of the holes is best in the plane (110) (in the direction <110>). Differences are explained by the band structures in these planes and also depend on the confinement of the carriers [25]. These considerations are important for nonplanar transistor architectures of FinFET type as discussed in Chapter 4. Other studies have also evaluated the change of direction of the channel by simple substrate rotation, in the <100> direction (at 45° to the usual direction <110>) [26–28].

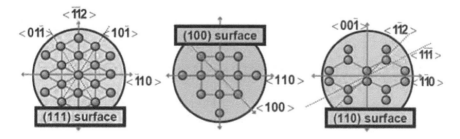

FIGURE 5.2 Network view and crystallographic orientations in the different silicon planes [22].

The "technology boosters," in particular, strain engineering, also depends on the considerations of crystallographic orientations. The results obtained in the standard cases are therefore valid only in one case and not necessarily in another configuration. As we have seen, bulk MOSFET technology has reached its limits, especially for electrostatic control. The improvements proposed are no longer sufficient for sub-20 nm technologies, and this is why other architectures of transistors are being proposed and studied. These innovative solutions are described in the following paragraphs.

As we have seen earlier, the architecture of the transistors was improved over the years to evolve toward the use of new gate materials as well as the introduction of mechanical stresses (by processes or by the addition of material at the level of the transistor). Although these many improvements have allowed bringing the technology used in bulk silicon to the 20 nm technological node [29–30], they will not be sufficient to respect the specifications of the technologies to satisfy the performance (I_{on}, I_{off}), especially gate control for dimensions < 20 nm [31–32].

5.2   FINFETS: SCALING AND DESIGN ISSUES

There are two (industrially viable) solutions for reducing leakage currents of the bulk structure. The first involves the use of thin-film structure: MOSFET transistors are fabricated on a Si substrate isolated from the bulk substrate by a layer of SiO_2 (called buried oxide). Thanks to a very thin Si thickness (<10 nm), the entire channel is controlled by the gate field, which improves the electrostatic control. The simulations show that the leakage current is even smaller than the film is the end [33]. It has therefore been estimated that the gate length must not be greater than four times the film thickness to maintain good control over the channel. We can also add that a thin film allows the formation of steep junctions, which allows better control of the electrical length. The ITRS anticipates that bulk technology will be replaced based on SOI. These architectures make it possible to better control SCE, thanks to the reduction of silicon thickness, allowing both a reduction of leakage current in the substrate, very good control of the gate on the channel and finally a modulation of the threshold voltage of the transistors using substrate doping or the application of a voltage on the rear face. With all the complexity of large SOI substrate manufacturing processes, a first approach to evaluate the interest of such architectures has therefore been to create these devices by modifying existing manufacturing processes, to form a structure under the conduction channel. These processes have been developed for many years and are perfectly controlled, the definition of structure to very small dimensions is possible.

5.2.1   THIN FILM ARCHITECTURES: LOCALIZED-SOI

To evaluate the influence on the function of conventional devices of a structure on the backside, many processes have been developed. By its ease of integration, the so-called "SOI local" structure has been particularly studied. The integration is based first of all on an epitaxial process, allowing control thin film growth with very high precision: on bulk silicon, a layer of SiGe (<20 nm) is grown to avoid any relaxation of the lattice parameter of the SiGe alloy (source of crystalline defects), then a last layer of Si is via epitaxy. This last layer represents the conduction channel, and we understand the advantage of this technique since the thickness of this channel is very well controlled.

The integration of the device then follows the standard CMOS process until the addition of the gate and junctions. With the structure being supported by STI isolations, it is then possible to remove the SiGe by selective etching. These steps, therefore, allow the formation of a suspended Si channel, hence the name silicon on nothing (SON) [34]. The selectivity of materials to the etching process depends on their nature (here the Ge concentration) and also the thickness of the layers [35]. It allows very fine control of the step of the release of the Si channel to form the SON. It is then possible to perform deposition of insulating materials, oxide, or nitride, which will fill the cavity under the silicon channel to thereby isolate the latter of the substrate [36–38].

5.2.2 DOUBLE GATE DEVICES

With a conduction channel perfectly isolated from the substrate, the double gate architecture (often known as DG) has the advantage of reducing leakage currents and improving control electrostatic [39]. Beyond isolation, the second gate of this architecture allows creating a second conduction channel geometrically opposite to the first. It may be noted that at first order that the thickness controlled by each gate now corresponds to half the thickness of Si. A simplified model allows us to account for the interest of this architecture on the control electrostatic channel [40]. It can also be added that the density of current can also be doubled for the same gate voltage since we form two conduction channels for the same size.

$$SS = \frac{\varepsilon_{Si}}{\varepsilon_{ox}} \frac{t_{ox}}{L_{EL}} \frac{T_{Si}/2}{L_{EL}} \times \left(1 + \frac{(T_{Si}/2)^2}{L_{EL}^2} \right) \times \Phi_d \tag{5.5}$$

$$DIBL = \frac{\varepsilon_{Si}}{\varepsilon_{ox}} \frac{t_{ox}}{L_{EL}} \frac{T_{Si}/2}{L_{EL}} \times \left(1 + \frac{(T_{Si}/2)^2}{L_{EL}^2} \right) \times V_d \tag{5.6}$$

The difficulty in the manufacture of a double gate device is the alignment of the two gates, which directly influences the performance of the transistor. Here, the alignment directly depends on lithographic alignment (precision of the order of 10 nm). The technique developed through a self-alignment process in the sizing of the two gates, which influences the electrostatic control of the device [41]. Local SOI-type processes (see Figure 5.3) have enabled the manufacture of self-aligned double-gate devices and simultaneous control of dimension. It should also be noted that the substrate will have to be highly doped to avoid the formation of a parasitic channel under the lower gate (potentially poorly controlled by the gate, therefore likely to degrade the electrostatics). The conduction channel being deposited epitaxially, its thickness can be easily controlled [42]. The manufacture of a thin layer of

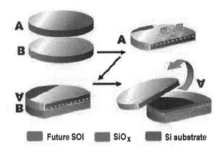

Future SOI   SiO$_x$ Si substrate

FIGURE 5.3   Local SOI formation process.

BOX (a few tens of nm) also allowed the development of variable conduction in the channel by the voltage present on the rear face of the substrate, which has the effect of modulating the threshold voltage of devices dynamically (unlike a substrate doping). In this sense, the substrate voltage across the insulator (BOX) equates to a second gate that will dynamically change the operation of the transistors.

The use of large-scale SOI substrates is now a reality (several suppliers are capable of producing SOI substrates, mainly SOITEC). STMicroelectronics and GlobalFoundries announced in June 2012, the production of 28 nm and 14 nm technology nodes [43]. The additional cost induced by the use of SOI substrates is quickly compensated by several manufacturing steps that could be avoided (doping, for example). This architecture also remains planar and is fully compatible with usual design rules already made on bulk technologies. With better control of the gate on the channel [44], better electrostatics in the device (compared to when the substrate is doped) is observed on the DIBL. This sensitivity to the rear face (the substrate located under the BOX) also allows to modulate the threshold voltage of the channel dynamically and maintain the proper functioning of an SRAM cell with a supply voltage of 0.4 V. BOX thickness also allows the channel to be more sensitive to the substrate while being electrically insulated. This is added to the modulation strategies of the V_T by the different work functions of the gate metals, already implemented in CMOS integration [45]. Finally, FDSOI technology has also been co-integrated with the bulk technologies to take advantage of planar thin-film structures dedicated to applications of low power (see Figure 5.4). The bulk substrate is indeed easily accessible

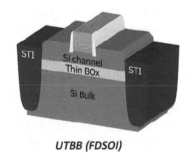

UTBB (FDSOI)

FIGURE 5.4   Schematic representation of an FDSOI device manufactured on SOI [22].

by selectively removing the BOX in selected areas [46], and the following processes allow simultaneous manufacture of both types of architectures.

A limit to using a very thin film is the increase in contact resistance for access to the channel. For this reason, FDSOI technology also includes a pitfall step, to increase the thickness of Si at the level of sources and drains. When properly controlled, this technique allows the use of different materials, such as SiGe or SiC. This technique is widely and even indispensable for the improvement of carrier transport in the transistors. The development of processes involving other materials and their compatibility of integration with other processes during manufacturing even allows the use of these materials in the transistor channel. For example, the channel manufactured using SiGe [47], which significantly improves performance by introducing stress in p-MOSFET as explained in Chapter 3. These processes are developed in such a way as to co-integrate these different materials into the same chip. It may be noted that this method is compatible with the miniaturization of the transistors since the performances are increased with the width of the transistor, which suggests that the mechanical stress is all the more so in small devices.

Thin film technology also allows for lower variability in bulk devices, thanks to the very good control of the processes [48–50], which makes it a very suitable technology for future nodes. The new technology boosters are essential for miniaturization and the reduction of the Si film thickness will help to improve the control of the channel by the gate. With dimensions < 10 nm today the variability control will be of the utmost importance at a large scale. Because of these physical limitations, other architectures are also envisaged to cope with the problems induced by miniaturization: the technologies for nonplanar multigate devices. To reduce leakage currents, control of the gate from all sides is desirable. The silicon layer must be thin enough for the gate to control the entire volume of Si. In this case, we may have several gates surrounding a channel in nonplanar transistors with multiple gates.

Like the so-called planar double gate architecture, there is an architecture double vertical gate, with a fin-shaped channel. The FinFET, proposed in 1998 (first n-FinFET [51] and then p-FinFET [52]). With the upper gate whose effect is neutralized by the addition of a hard mask (by a nitride), the FinFET can indeed be described as a DG device whose two gates and conduction channels are on the vertical planes. This allows the effect of overcoming the problem of self-alignment of the two gates encountered with a planar architecture [53]. FinFET has a higher conduction channel than dual-gate type conduction. For this, a hard mask is placed on the top of the Si channel before the gate deposition, so that the gate cannot have any effect on the upper plane. The conduction takes place only on the vertical faces of the fin, and so a double-gate device is perfectly aligned.

A FinFET having a high aspect ratio (large H and small W), presents the advantage of providing a large effective electrical width W_{eff} and therefore a strong current I_{ON}. Conduction is also done on both sides of dimensions H, whereas its footprint is minimal (reduced to W). The term was then derived as trigate to define a device having conduction on its three sides. In this case, the conduction by the upper surface contributes significantly in maintaining the threshold voltage at the short gate lengths (also called V_T roll-off) and improving the electrostatic control. At the same time, parasitic capacitances become less important than for a dual-gate FinFET [54, 55]),

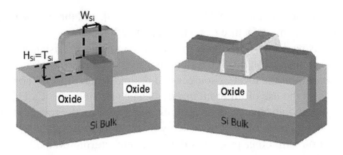

FIGURE 5.5   Diagram of a trigate FinFET on the bulk substrate as manufactured by Intel at the 22 nm technology node [22].

linked to the fact that the upper face becomes useful for current conduction. In this monograph, we shall mainly use the term TriGate for a device with three conduction surfaces, but with usual dimensions of the W ≥ H type (In general, the height H is fixed at about 10 nm and the width of the devices is, therefore, a minimum of 10 nm).

With the announcement of the launch by Intel of microprocessors based no longer on a conventional planar architecture, but 3D trigate FinFET on bulk Si (see Figure 5.5), the world of microelectronics changed. This architecture is called 3D because the conduction is no longer done on a silicon plane, but three faces. The difference from the double gate, the effective electrical width of the W_{eff} device will be defined by the height of the film H and its width W ($W_{eff} = W + 2 \times H$). These devices have the advantage of being made from conventional bulk silicon. The complexity of subsequent manufacturing steps is more of an argument in favor of cost reduction. But the FinFET bulk presents several difficulties: the complexity of lithography (over 100 nm high) and the insulation of the ends performed through a filling of oxide composites of the Fox (Flowable Oxide) type. To allow this filling of insulation without defects, the shape of the ends is not therefore not perfectly vertical but slightly triangular [56, 57]. It decreases electrostatic control (leakage in the wide zone, less well controlled by the gate) [58] but allows to maintain a good level of current, thanks to the gain in size (small footprint by W but large H so large, W_{eff}).

The Intel trigate FinFET is therefore perfectly suited for the manufacture of high-performance processors. It has nevertheless been shown that an imperfect end could cause a concentration of the current at its peak, and greater heating and variability effects. To change the level of current that will influence the circuit (in particular the actual current I_{EFF}), it is possible to change the W_{eff}. This allows ease in circuit manufacturing including different heights of Si, with a variability gain of 25% on an SRAM cell. In addition to very good electrostatic control and minimal footprint, nonplanar technologies also have the advantage of using the higher conduction in the silicon plane (110) since it is directly accessible on the fin. This allows additional benefit from the improvement of the transport of the holes and thus of better performance for p-MOSFETs. With the surface (100) under the gate being small, the effective carrier mobility for n-MOSFETs will be slightly lower. Nonplanar architectures have this peculiarity but are also compatible with the techniques of improvements developed for planar technologies such as the introduction of mechanical stress.

Finally, it may be noted that the FinFET, by design, is always linked to the substrate. To avoid current leakage in the OFF state, a zone is heavily doped for junction isolation. This process for the punch-through stopper is therefore technically difficult to implement [59–61] but necessary with a bulk substrate. For low-power applications, the SOI architecture is preferred and will present more of the advantage of facilitating the lithography of the Si.

5.2.3 FINFET AND TRIGATE FETS ON SOI

SOI multigate architectures have the same advantages as thin-film technologies. The presence of three gates allows excellent control electrostatic, which allows us to relax the T Si dimension in comparison with the structures on the thin film. Access to lateral conduction channels, on the plane's crystallographic data of Si (110) on standard plates, also allows for effective holes improved in comparison to a planar architecture, always with a directed channel in the <100> orientation.

The use of an SOI substrate brings several advantages to the manufacture of nonplanar devices. First of all, the isolation of the transistor is total, thanks to the BOX. Punch-through techniques are useless here, so the process is simpler technically but also less expensive in an industrial context. The process of engraving is also facilitated since the BOX will serve as a barrier layer (see Figure 5.6). A schematic diagram of the crosssection for different kinds of gate architectures for FinFET is shown in Figure 5.7. The dimensional control is then improved and the variability is reduced. This aspect is all the more important as the devices will generally be used in high-speed circuits. For this, it is necessary to control the so-called spacer patterning or sidewall image transfer (SIT). The technique SIT [62] makes it possible to obtain structures whose dimensions are not attainable by current lithography methods [63]. The principle is based on the division of dimensions since we reuse two spacers formed around a structure named mandrel as a mask. This has been already used in an industrial context on nonplanar transistors [56]. Optimizing this process has a significant impact on the performances of the transistors [63].

With a relatively small spacing, SiGe source/drain type stressors have little space but the residual state of the strain remains in one case or the other [64]. Finally, unlike planar technologies and, in particular, the thin-film technologies, the FinFET does not offer multiple threshold voltages. It is interesting to find applications where the

FIGURE 5.6 Schematic representation of a FinFET architecture on SOI [22].

performance needs both multiple threshold voltage and low power are necessary. Different V_T is available by the use of several gate metals, as in a planar technology [65, 66]. The possibility of having different threshold voltages, by doping the substrate or by backside potential is possible only for a device having a silicon thickness small.

For FinFETs with height H generally of the same order as its width W (a TriGate device), the three gates are substantially identical in dimensions, with a $W \ll H$, which is rather a double vertical gate, while the TriGate is more relaxed and therefore less difficult to manufacture. Although the silicon film height is generally greater than 10 nm, we can find some resemblance to the planar FDSOI, but whose miniaturized dimension is its width W, rather than the T_{Si} film thickness. TriGate is more like an architecture intermediate and benefits from these two structures. Moreover, this makes it a perfect candidate for the evolution of technology FDSOI technology nodes for ultimate scaling (<10 nm). Several authors have reported TriGate devices on SOI having conduction channels on a vertical plane (110) which is advantageous for the p-MOSFET with widths from 60 nm [67] down to 20 nm [58] and even 10 nm, thanks to the SIT technique. Like the FinFET, the TriGate shows excellent electrostatic control and particularly when its width W is decreased. A DIBL of <100 mV/V when $W < 15$ nm and for a gate length of $L = 25$ nm [68].

It has been shown that the manufacture of trigate is compatible with 3D strain techniques or stress memorization techniques (SMT) [69]. Transistor manufacturing TriGate with a SiGe channel has also been demonstrated to improve the performance of p-MOSFET devices [70, 71]. The TriGate also benefits from a similar advantage to film technology thin, since its development is carried out on a thin BOX structure: the possibility of dynamically modifying the threshold voltage of the device. Finally, as for the FinFET, the SIT technique allows a dense integration of TriGate. Studies on these forms of TriGate are indicative of very good electrostatic since the control of the gate is optimal [72]. The decrease in gate lengths and the need for density have allowed the appearance of triple-gate technologies by Intel from the 22 nm node [56]. The needs of advanced technological nodes are not achievable by bulk technology. Also, other manufacturers choose to continue the planar integration through the thin film FDSOI, which is expected to further push Moore's law to the 10 nm node. The new 3D architectures will nevertheless be necessary for better control of the devices when the gate length reduces further to sub-10 nm. To meet the needs of the MOSFETs, the possibilities of integrating and compatibility with technology boosters such as mechanical stress should be studied in detail in this type of architecture.

5.3 VIRTUAL FABRICATION OF BULK-SI TRIGATE FINFETS

Semiconductor device manufacturing usually involves hundreds of processing steps, months of processing time, and unpredictable relationships between the tool performances and yield. Because of the thermal coefficient of expansion mismatch, stresses in electronic devices may cause not only premature mechanical failures but also to affect the function of the semiconductor devices. The proliferation of TCAD tools has a significant economic impact on semiconductor product development. The effective use of TCAD tools saves experimentation time for calibrating the process, device parameter extraction, and minimizes the number of trial-and-error iterations. TCAD tools provide a more comprehensive approach to characterize technologies and help to optimize the performance [73] as a complement to experimental trials.

As discussed above, multigate FET technology is the best alternative that can extend scaling to the sub-10 nm technology nodes with the minimum additional processing costs. From the fabrication perspective, the most likely candidate for wide adoption among the multigate devices is the FinFETs. FinFETs are broadly classified into two categories based on the type of substrate, that is, on SOI [74, 75] and bulk-Si [76, 77]. The process steps of bulk FinFET are different from planner MOSFETs. Though FinFET process steps have been reported by many research groups [78, 79], detailed fabrication process steps are not available in the public domain. Process steps play an important role in determining device performance.

In this section, TCAD simulations (process and device) are performed for trigate FinFETs. Process steps considered are for 2D and 3D simulations. Such considerations will be verified from the electrical characterization using device simulations with suitable transport models as discussed in Chapter 2. To evaluate the device performance, we have analyzed ballistic and nonballistic transport models considering the quantum mechanical effects. The energy balance model and the Drift-Diffusion model are also compared along with the incorporation of necessary quantum correction.

In this section, we discuss the detailed process steps for the fabrication of n-channel FinFET. The basic process steps include deposition, oxidation, ion implantation, lithography, and etching which are used to grow the fin, field oxide, gate oxide, polysilicon, and metal contacts of the FinFETs. Typical FinFET gate structures are shown in Figure 5.7. The vertically thin channel is formed on the side vertical surface of the Si-fin, and the current flows in parallel to the wafer surface.

3D process simulation is performed using the VictoryProcess simulator [80]. We can optimize 3D process simulation performance by using a proper model for each

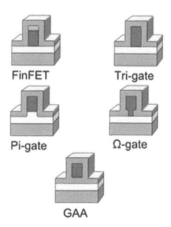

FIGURE 5.7 Schematic diagram of the cross section for different kinds of gate architectures for FinFET.

processing step. At first, the 3D symmetric mesh is created. Here, we have performed a mixed dimension simulation, in which we can define both 3D and 2D simulation domain, but to start with 2D VictoryProcess symmetry is used by setting the parameter FLOW.DIM (=2D) to a 2D mode of simulation.

Following the initialization, we have defined the volume data mesh. After the volume data specification, the fin (active area) was specified and doping to the fin was introduced. We have simulated all these steps in the 2D mode because the mask layer which defines the fin is made symmetric in the y-direction (all mask edges within the simulation domain are parallel to the y-axis). We can obtain a significant simulation performance gain by using the 2D mode for this processing sequence. This is because implantation and diffusion simulation steps used are performed fully in 2D. The main reason is the short simulation time required in 2D mode. The full 3D volume data mesh that we have defined contains $31 \times 29 \times 70$ mesh planes and 62,930 mesh points, while the volume data mesh used by the 2D mode only contains 31×70 mesh lines and 2,170 volume data mesh points. It is also possible to obtain significantly shorter run times for oxidation process steps when we use the 2D mode.

The main FinFET process flow is as follows:

A. Fin formation
B. Gate patterned
C. Spacer formation
D. Source/drain formation
E. S/D I_{on} implantation and contact formation.

5.3.1 FIN FORMATION

To start with the simulation, initially, the volume information is specified as discussed above. After the volume data specification, the fin (active area) is defined and followed by the fin doping. Initially, the silicon substrate layer is grown on the Si wafer. A dry etching step with geometrical models is formed to transfer the masks into the structure. A 50 nm followed by a 250 nm deep etching process is used to optimally transfer the mask pattern into the silicon substrate. Oxidation followed by etching steps is adopted to prepare the substrate for the fin formation and trench isolation with the help of a suitable mask. The fin height formed is 50 nm and the fin width is 30 nm. After the formation of the fin, it is doped with boron with a dose of 1×10^{14} cm^{-2} with an energy of 1 keV.

5.3.2 GATE FORMATION

Once the fin is created, it is covered with the gate material that is nitride and polysilicon. These two deposition steps in 2D mode are possible and no mask is required. Since this sequence only uses deposition steps, the 2D mode does not provide any performance gain. As the patterning step to form the gate uses a mask ("Poly," as shown in Figure 5.8), which no longer satisfies the 2D geometrical symmetry condition, we have to switch over from the 2D mode to the 3D mode before we can perform the next processing step. Figure 5.9a shows the results of the Si_3N_4 deposited on

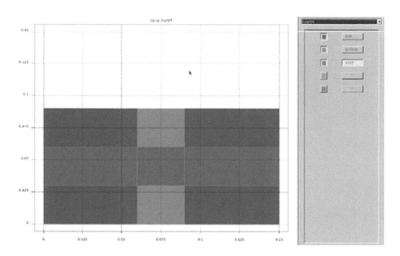

FIGURE 5.8 Mask layers that have been used so far in the process flow. Mask layer "Poly" (purple) is on top [80].

top of the fin. Figure 5.9b shows polysilicon deposited on top of the nitride layer. To start with the 3D mode process steps, the SIMULATION MODE has to be set first. SIMULATION MODE FLOW.DIM=3D. The FLOW.DIM parameter selects the simulation mode. It produces the 3D structure as shown in Figure 5.9a. To finalize the formation of the gate, etch the polysilicon with the mask defined by the mask layer "Poly" (see Figure 5.8). The structure after 30 nm long gate formation is shown in Figure 5.9c.

5.3.3 Spacer Formation

After forming the gate, we need to create the spacer. This is done by the deposition of oxide followed by etching. Figure 5.9d shows a thick oxide layer deposition. By selective etching gate polysilicon region, the oxide was etched out and open for gate contact formation as shown in Figure 5.9e. Similarly, the oxide and nitride etching has been performed to expose the silicon area which will be used for the formation of source and drain on either side of the fin. Hence, the oxide spacer layer was formed to isolate the fin from the source and drain regions of the fin which can be seen in Figure 5.9f.

5.3.4 Formation of Source/Drain

The source and drain contacts are formed by covering the source and the drain regions with a layer of polysilicon. Before the source–drain contact formation, the photoresist material is deposited and patterned in such a way that it covers the spacers as shown in Figure 5.9g. Figure 5.9h shows the formation source and drain contacts by polysilicon deposition. Finally, the etching process leaves the device ready for ion implantation as shown in Figure 5.9i. Figure 5.9g–i shows the intermediate

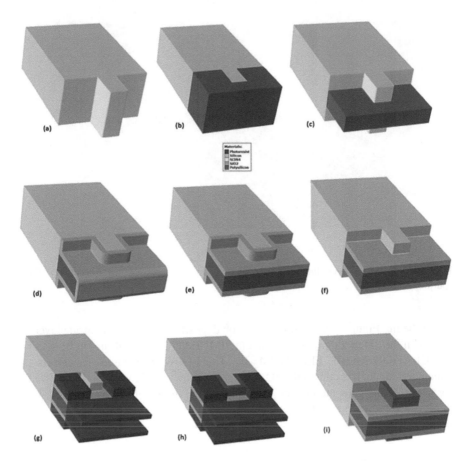

FIGURE 5.9 (a) Si fin after nitride disposition (b) 3D view of the polysilicon layer deposited on top of the fin (c) The FinFET structure after gate Formation (Lg = 30 nm). (d) Oxide deposition on the top surface (e) Etching of oxide for gate contact formation (f) Removal of oxide layer from the side of the fin and etching of nitride layer exposes Si area for S/D formation (g) Photoresist deposition and selective etching (h) Polysilicon deposition on the source and drain for contact formation. (i) Removal of photoresist form source and drain spacer region.

results obtained by this processing sequence. To complete the processing part, one needs to follow process steps; the source–drain doping by implantation and activate it with an annealing step as will be described next.

5.3.5 S/D Ion Implantation and Contact Formation

The implant simulation step performs two ion implantation steps. In this case, the first one with an ion beam rotation of 0° and the second one with a rotation of 180°. The extension implantation is performed with arsenic ions at a dose of 10 keV to form the source/drain doping. To complete this process simulation sequence, one needs to apply an annealing step to activate the dopants. The diffusion model within the

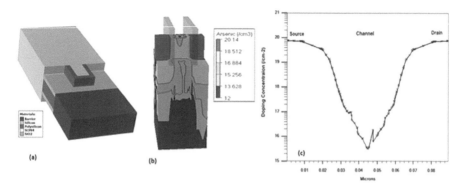

FIGURE 5.10 (a) Illustration of barrier formation for angle implant for S/D doping (b) Arsenic Doping Profile after ion implantation (c) Doping Profile in the fin from Source to Drain.

material silicon performs the redistribution of the dopants during this high-temperature processing step. Figure 5.10a shows simulation results created by this processing sequence when the device was prepared for ion implantation. The doping profile obtained from process simulation has been shown in Figure 5.10b. Figure 5.10c shows the quantitative doping values across the fin from source to drain. The structure shown in Figure 5.11 is the final FinFET device structure obtained from VictoryProcess. This structure is exported for device simulation as will be discussed in the next sections. The geometry detail of the fabricated FinFET is shown in Table 5.1.

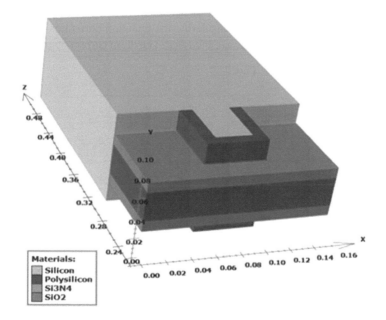

FIGURE 5.11 Final FinFET device structure was obtained from the simulation.

TABLE 5.1

Geometrical Detail of the Virtually Fabricated Device is Shown in Figure 5.11

Geometrical Parameters	Value
Gate Length	30 nm
Fin Height	50 nm
Fin Width	30 nm
S/D Doping	1e20
Substrate Doping	1e15
Gate Oxide Thickness	1 nm

5.3.6 Electrical Performance

Once the FinFET is fabricated using the process steps such as etch, deposit, diffusion, and Monte Carlo implantation modules of VictoryProcess, the structure is saved using a full 3D Delaunay mesh and is then transferred to VictoryDevice [81] for device simulation. But before we go for device simulation, we need to decide on the suitable models (as discussed in Chapter 2) to be used for such nanoscale devices. In the simulation, the ATLAS device simulator [82] was used for the FinFET simulation in which various quantum corrected models are implemented. In device simulations, we have used the Bohm quantum potential (BQP), density gradient (DG), energy balance (EB), and Drift-Diffusion (DD) models to evaluate their suitability for nanoscale devices.

Figure 5.12 shows the transfer characteristics for the FinFET considering transport models viz., BQP, energy balance, and Drift-Diffusion models. The drain current is highest in the EB model. Considering the quantum corrections (in the BQP model), the energy balance models are modified. The drain current obtained is shown

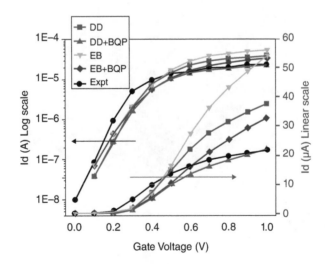

FIGURE 5.12 Transfer characteristics of p-FinFETs using different transport models.

TABLE 5.2
Extracted Subthreshold Parameters of the Virtually Fabricated p-FinFET Simulated with Different Models

Parameters/Models	DD	BQP	EB	EB+BQP
V_{TI}	0.290697	0.257007	0.331438	0.280831
β	0.00162548	0.000968191	0.0022832	0.00110013
Θ	0.753201	0.858565	0.619572	0.289235
SS	0.115959	0.117462	0.120416	0.121369
I_{on} (Amp)	3.7574e-005	2.19597e-005	5.39682e-005	3.27473e-005
I_{off} (Amp)	3.88238e-008	4.10358e-008	7.13382e-008	7.44315e-008

as a blue line (Energy Balance + BQP). The reduction of current from only the EB model is due to the quantum confinement of some of the carriers (out of total carriers) that take part in current conduction. To compare with the nonballistic transportation model, I_d-V_g characteristics were obtained using the Drift-Diffusion model along with the quantum correction (with BQP) model. As the Drift-Diffusion model considers the scattering effect, it is expected that it should produce a lower current than that of the energy balance model. The Drift-Diffusion model with quantum mechanical effect also performs in the same way as in the case of the EB model and it reduces the value of current due quantum confinement.

The subthreshold conduction parameters for the above FinFET using different transport models are extracted and listed in Table 5.2. The threshold voltage is found by taking the x-intercept of the maximum slope to the I_d-V_g characteristics and subtracting half the drain voltage. The gain (or Beta) is defined as the value of the steepest slope to the I_d-V_g characteristics and divided by the drain voltage. The mobility roll-off parameter (or Theta) is also extracted.

The I_d-V_d characteristics have also been obtained for different transport models and are shown in Figure 5.13. For different transportation models, I_d-V_d curves follow the same trend as transfer characteristics. The increase in drain current with the ballistic model is because of reduced scattering and output impendence. The i_d is highest with the energy balance model followed by drift-diffusion without quantum effect. When the quantum correction is included, the drain current decreases by ~40% compared to the energy balance model and by ~50% with the Drift-Diffusion model. While the EB model and DD model are compared, the EB model shows a ~35% and ~44% increment of current compared to the DD model with and without quantum corrections, respectively.

5.4 STRESS TUNING USING EPI-SIGE SOURCE/DRAIN STRESSOR

The leading microelectronics manufacturers recognize the need to include strain in their processes which is known as "technology booster." This is because the targets set by IRDS cannot be achieved by geometry scaling alone. Techniques for the introduction of intentional strain in CMOS technologies have been considered in detail in Chapter 3. As such in planar technologies, epitaxially grown SiGe in the S/D regions

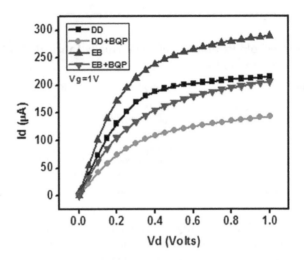

FIGURE 5.13   Output characteristics for different transportation models.

enhance the performance of the FinFET by imparting strain to the channel and forming sharp ultra-shallow junctions [83]. Performance improvement depends on several
other factors:

(1) Distance between S/D,
(2) Raised or embedded S/D growth,
(3) Epi growth profile,
(4) S/D etch shape profile,
(5) S/D etch depth, and
(6) The volume of Ge material for p-MOSFET.

The distance between the source and drain regions is primarily determined by the
S/D junction formation and is thus dependent on the type of epi formation as well:
raised S/D and/or embedded S/D greatly influence the device performance [84]. The
raised S/D generates biaxial strain while the embedded S/D generates uniaxial strain
to the channel where the biaxial strain is less than the uniaxial strain. It is possible
to obtain significant carrier mobility enhancement for most of the channel length,
using remotely located epitaxial stressor material in the source and drain regions.
The SiGe source/drain is going to be the main source of stress below the 14 nm
p-FinFET due to the tight gate pitch and the gate-last high-k metal gate (HKMG)
process [85]. The design of epitaxial source–drain stressor is possible in device
engineering options [86] and the key scaling enablers. In this regard, Tan et al. [87]
reported a p-channel strained FinFET with a SiGe S/D [88]. The authors have investigated the selectively grown $Si_{1-x}Ge_x$ ($x = 0.33$–0.35) with a boron concentration of
1×10^{20} cm^{-3} which was used to elevate the S/D regions of bulk FinFETs in the
14 nm technology node. In the next section, the process simulation for FinFET with
epitaxially grown source/drain has been presented for extremely scaled transistors at
a 7 nm technology node.

To design FinFET with an epitaxially grown source and drain with SiGe, we need to follow different process steps. The main process steps for p-FinFET are stated below:

- Si Substrate
- Strain-relaxed buffer growth (virtual substrate)
- Fin formation and V_T implant adjustment
- Poly gate deposition and RIE
- Equal spacer formation
- Source–drain epitaxy with SiGe
- HKMG formation and recess
- Self-aligned contact (SAC) formation

5.4.1 FIN FORMATION

The fin formation process is similar to the process steps discussed before except it is on the strain relaxed buffer as shown in Figure 5.14a–c. Initially, the silicon substrate layer is grown on the Si wafer. A dry etching step with geometrical models is carried

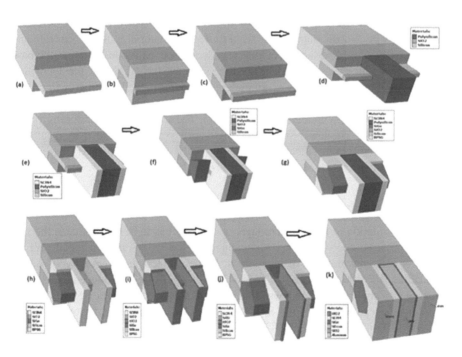

FIGURE 5.14 (a) Fin Formation, and (b) trench isolation with SiO_2. (c) Active area formation (d) Formation of the polysilicon layer (e) Nitride spacer formation (f) Formation of SiGe epitaxial layer (SiGe growth with initial 120° angle). (g) Diamond shape S/D stressor after epitaxial layer growth (h) Formation of the gate oxide (i) HfO_2 deposition (j) Etching of HfO_2. (k) Full device after metal contact.

out to transfer the masks into the structure. A 35 nm deep etching process at an angle of 87° is used to optimally transfer the mask pattern on the silicon substrate. It forms the tapered strain relaxed buffer layer with a base width of 10 nm and a top width of 7 nm. Oxidation followed by etching steps is performed to prepare the substrate for fin formation and trench isolation with the help of a suitable mask. The resulted fin height is 20 nm and the fin width is 7 nm shown. Fin is doped with a boron implant with a doping concentration of 1×10^{15} cm^{-3} to form a p-channel device which is also a vital part to set the threshold voltage of the device.

5.4.2   POLYGATE DEPOSITION AND SPACER FORMATION

Once the fin is formed, it is covered with the polysilicon gate material and selectively etched to form the gate area. The polysilicon is deposited and etched away with a suitable mask to cover the gate area. The nitride layer is deposited and etched to form the spacer layer on both sides of poly. The poly and nitride layer deposition steps are shown in Figures 5.14d and e.

5.4.3   SOURCE–DRAIN EPITAXY

One of the most important processes in the semiconductor industry is the epitaxy. Normally conformal epitaxial growth over 3D structures is performed using selective deposition. The deposit statement is used to deposit material with a given rate and ratio over all exposed surfaces. For selective deposition, Si is considered as the base material over which we want to deposit SiGe. Then, we have to invoke the deposition after stating the simulator SiGe as epi material for deposition. The epi growth is controlled with an initial epi angle of 120° up to vertical 21 nm height and then 60° from the middle of the epi with a thickness of 21 nm. The source and drain doping values are 1×10^{20} cm^{-3}. The lower side of the SiGe epi layer is surrounded by insulating material borophosphoro silicate glass (BPSG). The simulated epitaxial growth of SiGe on top of silicon is shown in Figure 5.14f–h.

5.4.4   HKMG FORMATION AND CONTACTS

After the formation of the SiGe epi layer, gate oxide has to be deposited. In the fabrication process, one of the major challenges is the work function metal (WFM) recess associated with the SAC integration. As the gate length is aggressively scaled, controlling WFM recess becomes very challenging. HfO$_2$ is introduced as a high-k material at 7 nm node in such a way that it can reduce V_T dependency on gate length as high-k is separated from the SAC cap and fully encapsulated by metal, which prevents OH diffusion into the HfO$_2$. In the simulation, initially, 0.8 nm thick SiO$_2$ is deposited followed by HfO$_2$ which has been used as the high-k gate material. The thickness of the HfO$_2$ is 0.16 nm. The effective gate oxide thickness is maintained at ~1 nm. Finally, contacts are made for source–drain and gate with Al. The detailed steps are shown in Figure 5.14i–k. The final structure has been shown in Figure 5.14k. The front and cross-sectional views have been compared with the reported device by Intel [56] as shown in Figures 5.15.

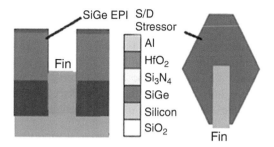

FIGURE 5.15   p-channel FinFET structure is used in the simulation. SiGe epitaxy in the S/D region is shown (side view of the simulated device). Fin with epitaxial S/D stressor used for simulation.

5.4.5   STRESS TUNING

The stress analysis is performed by the stress analysis tool VictoryStress [89] which determines the stress in various regions of the fin in the device whose fabrication steps are discussed above. The lattice mismatch between Si and $Si_{1-x}Ge_x$ generates uniaxial compressive stress in the longitudinal (gate length) direction, which improves hole mobility [25]. Figure 5.16 shows the schematics of the

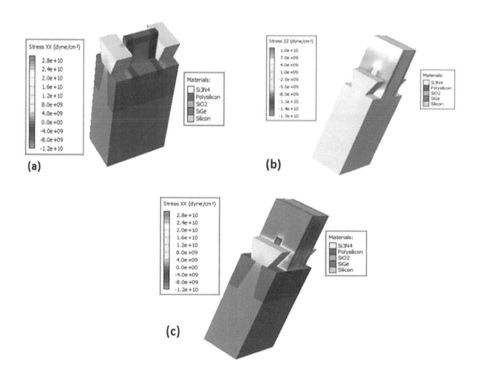

FIGURE 5.16   Stress mapping in the device structures: (a) after SiGe -epi, (b) and (c) after spacer.

process-simulated device structure from different stages of the process simulation, together with the calculated strain distribution profiles. The stress map in the xx- and zz-directions are shown in Figures 5.16b and 5.16c. All the maps show strain in the Si fin. Figure 5.16c shows the fin embedded in SiGe resulting in a net compression of the channel area between the SiGe source/drain areas. Whereas the strain distribution in the horizontal direction, xx, is directly visible, the compressive strain under the SiGe could not be visualized, likely due to the presence of the bulk silicon substrate.

5.4.6 Stress Analysis for Optimization of Source/Drain Stressor

Figures 5.17–5.19 show the SiGe ($x = 0.5$) source–drain stressor geometry variation (see shapes for the green region). In the process simulation, reactive ion etching (RIE) was performed for various process conditions to achieve these source–drain stressor structures.

Using TCAD simulations, we can map the stress and calculate the orientation-dependent mobility on sidewalls of trigate FinFETs [90–93]. So far, stress distribution has been shown considering the 3D profile. Figure 5.20 shows the 2D and 1D stress distribution map in the fin. Due to a change in the shape of the stressor, the strain/stress transferred to the fin is different which ultimately affects the electrical performance of the device. As an example, we have simulated the device using VictoryDevice [81] which will be described in the next section.

For device simulation, the basic models such as models for mobility, carrier recombination, and tunneling effects have been included in the simulation. Typical physics-based models such as concentration and field-dependent model for mobility, Shockley–Read–Hall (SRH) recombination model, the nonlocal band-to-band

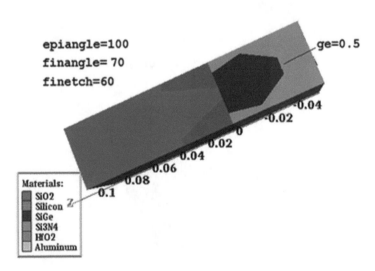

FIGURE 5.17 Schematic of the S/D stressor structure after SiGe -epi. See inset for RIE process conditions-1.

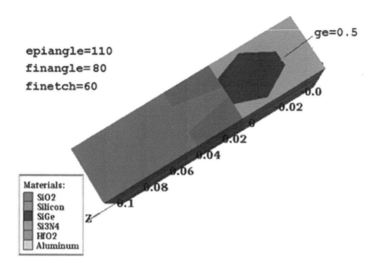

FIGURE 5.18 Schematic of the S/D stressor structure after SiGe -epi. See inset for RIE process conditions-2.

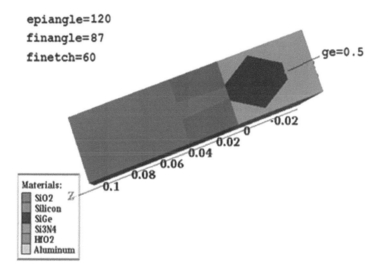

FIGURE 5.19 Schematic of the S/D stressor structure after SiGe -epi. See inset for RIE process conditions-3.

tunneling model, and bandgap narrowing model, and Fermi–Dirac statistics model are implemented in simulation.

5.5 ELECTRICAL PERFORMANCE

In the device simulation, we calculate first the unsaturated I_d-V_g characteristics without invoking the strain enhancement. In the next device simulation, we take the strain

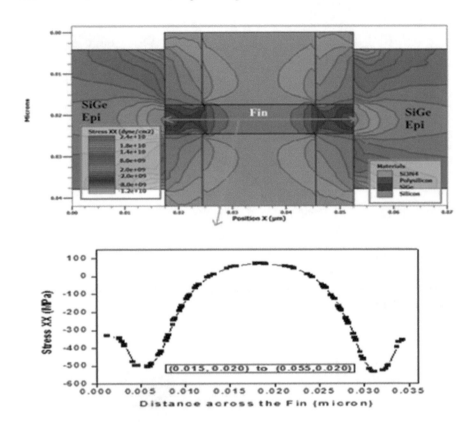

FIGURE 5.20 Typical stress map in the fin section after SiGe -epi.

into account. The difference in the I_d-V_g characteristics due to the strain enhanced mobility is shown in Figure 5.21. Table 5.3 shows the critical device parameters obtained for the above no strain and strain enhanced cases for 7 nm p-channel FinFET.

Figure 5.22 shows the transfer characteristics (I_d-V_g) of the FinFET device with four different source–drain epi structures (see Figures 5.17–5.19). It is interesting to note that the structure shown in Figure 5.18 gives rise to the highest drain current. However, when the geometry of the stressor is changed (see Figure 5.19), the drain current reduces, thus showing the sensitivity of source–drain stressor geometry on electrical performances. Device parameters (threshold voltage, subthreshold slope, I_{on}, and I_{off}) are extracted and are shown in Table 5.4 for different shapes of the source/drain epi stressor.

5.5.1 SUMMARY

In this chapter, the evolution of CMOS technology has been discussed briefly. Many modifications of the architecture of the transistor have made it possible to reach advanced nodes (22/10 nm). The planar devices, of the localized thin film or double

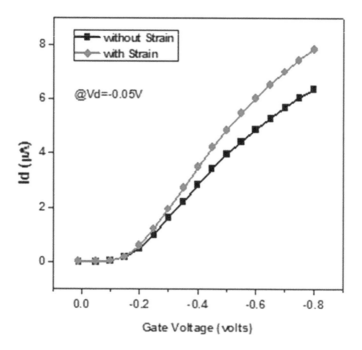

FIGURE 5.21 I_d-V_g characteristics for of 7 nm p-FinFET (a) with and (b) without strain.

TABLE 5.3
Extracted Parameters for the 7 nm p-FinFET With and Without Stress

Device	No Stress	With Stress
V_T (V)	0.14731	0.1499
Beta	0.000250726	0.000310318
Theta	0.439162	0.43871
Sub-V_T (mV/dec)	70.8052	70.4123
I_{on} (μA)	6.35946	7.84846
I_{off} (nA)	0.200631	0.220533

gate type devices, have been studied for their ease of integration with the design rules of bulk. The first nonplanar transistors of the type Trigate FinFET on bulk appeared in 2011 in mass production at the 22 nm node. The miniaturization of such devices is, however, tricky since their dimensions reach the limits of the processes. The bulk substrate seems to get limited for low-power applications. The use of compressive stress and strain shows possible improvements in the performances of the transistors. FinFET full virtual fabrication process is shown. For process optimization, suitable models have been adopted for oxidation, lithography, and ion implantation. After process simulation, the device simulation has been performed with ballistic and non-ballistic transportation mode with consideration of a quantum mechanical effect. The energy balance model has been used to account for the effects of ballistic carrier

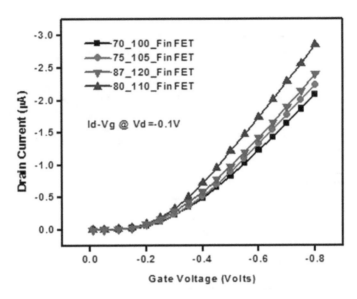

FIGURE 5.22 Transfer characteristics (I_d-V_g) of the FinFET device with four different source–drain epi structures (see Figures 5.17–5.19).

TABLE 5.4
Extracted Parameters for FinFET with Different SiGe Stressor Shapes

Device Parameters	Stressor Types (Fin-angle_Epi-angle)			
	70_100	75_105	80_110	87_120
V_T (V)	0.309023	0.298515	0.276981	0.302709
Beta	8.95727e-005	9.38376e-005	0.000114434	0.000101549
Theta	0.109273	0.104625	0.0959785	0.106448
Sub-V_T (mV/dec)	86.5275	94.6067	93.0587	94.2563
I_{on} (µA)	2.08694	2.23561	2.84952	2.39803
I_{off} (nA)	0.6572	0.9181	0.1419	0.12602

transport which significantly increases the drain current. The increase in drain current with the ballistic model is because of the decrease in the output impedance. However, when the effect of quantum confinement is taken into account, the drain current decreases significantly. The BQP model is used to introduce quantum confinement in 3D. The drain current shows a 44% increment with the EB model in comparison to the DD model. But with consideration of the quantum effect, the drain current decreases up to 40% with the energy balance model and 50% with the Drift-Diffusion model. The critical device parameters such as threshold voltage, subthreshold slope, I_{on}, and I_{off} have been extracted.

We have examined strain distribution in FinFET structures with epitaxial $Si_{1-x}Ge_x$ stressors deposited around the source/drain region. The SiGe source–drain stressor has been used to generate compressive stress in the channel. The linear elasticity theory has been implemented in simulation to describe the evolution of the stress for various stages of the process. We have been able to predict the possibility of tuning the residual stress by changing the geometry of the SiGe stressor (shape) during epitaxial growth. Also, the strain in the fin and channel is found to be very sensitive to such structural factors as the stressor thickness and fin width. The 2D strain variation data across the FinFET structure obtained in this study are particularly beneficial for device design for FinFET, offering full information on strain variation across the structure.

TCAD simulations of a trigate FinFET have been discussed in detail. An appropriate 3D TCAD modeling approach is proposed to handle the new FinFET-specific design and meet the process challenges. For the first time, stress profiling analysis is adopted to simulate the process of SiGe epitaxial growth in p-type FinFETs. We have shown using TCAD mechanical stress tuning techniques and electrical simulation that significant improvements in the performance of FinFET at the 7 nm technology node are possible. This study will undoubtedly provide a guideline for device design using stress tuning in the channel of trigate FinFETs at the 7 nm technology node and below.

REFERENCES

[1] IEEE, "International roadmap for devices and systems, 2020 update, beyond CMOS." 2020.

[2] B. Doris, B. Desalvo, K. Cheng, P. Morin, and M. Vinet, "Planar fully-depleted-silicon-on-insulator technologies: Toward the 28 nm node and beyond," *Solid. State. Electron.*, vol. 117, pp. 37–59, 2016, doi: 10.1016/j.sse.2015.11.006.

[3] K. Kuhn, "Chapter 1 - CMOS and beyond CMOS: Scaling challenges," *High Mobility Materials for CMOS Applications*. Woodhead Publishing, pp. 1–44, 2018, doi: doi:10.1016/B978-0-08-102061-6.00001-X.

[4] F. Balestra, "Challenges for high performance and very low power operation at the end of the Roadmap," *Solid. State. Electron.*, vol. 155, pp. 27–31, 2019, doi: doi:10.1016/j.sse.2019.03.011.

[5] J. P. Colinge and A. Chandrakasan, *FinFETs and other multi-gate transistors.* Springer US, 2008.

[6] T. Skotnicki, J. A. Hutchby, T. J. King, H. S. P. Wong, and F. Boeuf, "The end of CMOS scaling: Toward the introduction of new materials and structural changes to improve MOSFET performance," *IEEE Circuits Devices Mag.*, vol. 21, no. 1, pp. 16–26, 2005, doi: 10.1109/MCD.2005.1388765.

[7] N. Horiguchi, T. Usuki, K. Goto, T. Futatsugi, T. Sugii, and N. Yokoyama, "*Direct Tunneling Memory (DTM) utilizing novel floating gate structure,*" in *Technical Digest - International Electron Devices Meeting*, 1999, pp. 922–924, doi: 10.1109/iedm.1999.824299.

[8] M. A. der Maur, M. Povolotskyi, F. Sacconi, A. Pecchia, and A. Di Carlo, "Multiscale simulation of MOS systems based on high-κ oxides," *J. Comput. Electron.*, vol. 7, no. 3, pp. 398–402, 2008, doi: 10.1007/s10825-007-0160-8.

[9] G. S. Lujan et al., "*A new method to calculate leakage current and its applications for sub-45nm MOSFETs,*" in *Proceedings of 35th European Solid-State Device Research Conference, 2005. ESSDERC 2005.*, 2005, pp. 489–492.

[10] G. S. Lujan, S. Kubicek, S. De Gendt, M. Heyns, W. Magnus, and K. De Meyer, "*Mobility degradation in high-k transistors: The role of the charge scattering,*" in *European Solid-State Device Research Conference*, 2003, pp. 399–402, doi: 10.1109/ESSDERC.2003.1256898.

[11] J. F. Yang, Z. L. Xia, G. Du, X. Y. Liu, R. Q. Han, and J. F. Kang, "*Coulomb Scattering induced mobility degradation in Ultrathin-body SOI MOSFETs with high-k gate stack,*" in *ICSICT-2006: 2006 8th International Conference on Solid-State and Integrated Circuit Technology, Proceedings*, 2006, pp. 1315–1317, doi: 10.1109/ICSICT.2006. 306146.

[12] C. H. Jan et al., "A 32nm SoC platform technology with 2nd generation high-k/metal gate transistors optimized for ultra low power, high performance, and high density product applications," 2009, doi: 10.1109/IEDM.2009.5424258.

[13] S. E. Thompson et al., "A 90-nm logic technology featuring strained-silicon," *IEEE Trans. Electron Devices*, vol. 51, no. 11, pp. 1790–1797, Nov. 2004, doi: 10.1109/TED. 2004.836648.

[14] C. H. Chen et al., "*Stress Memorization Technique (SMT) by Selectively strained-nitride capping for sub-65nm high-performance strained-Si device application,*" in *IEEE VLSI Tech. Symp. Dig.*, 2004, pp. 56–57, doi: 10.1109/vlsit.2004.1345390.

[15] K. J. Kuhn, "Considerations for ultimate CMOS scaling," *IEEE Trans. Electron Devices*, vol. 59, no. 7. pp. 1813–1828, 2012, doi: 10.1109/TED.2012.2193129.

[16] K. Ang, J. Lin, C. Tung, N. Balasubramanian, G. Samudra, and Y. Yeo, "*Beneath-The-Channel Strain-Transfer-Structure (STS) and embedded source/drain stressors for strain and performance enhancement of nanoscale MOSFETs,*" in *2007 IEEE Symposium on VLSI Technology*, 2007, pp. 42–43, doi: 10.1109/VLSIT.2007.4339719.

[17] K. W. Ang, J. Lin, C. H. Tung, N. Balasubramanian, G. S. Samudra, and Y. C. Yeo, "Strained n-MOSFET with embedded source/drain stressors and Strain-Transfer Structure (STS) for enhanced transistor performance," *IEEE Trans. Electron Devices*, vol. 55, no. 3, pp. 850–857, 2008, doi: 10.1109/TED.2007.915053.

[18] T. Liow et al., "N-Channel (110)-sidewall strained FinFETs with silicon-carbon source and drain stressors and tensile capping layer," *IEEE Electron Device Lett.*, vol. 28, no. 11, pp. 1014–1017, 2007, doi: 10.1109/LED.2007.908495.

[19] J. Bardeen and W. H. Brattain, "Physical principles involved in transistor action," *Phys. Rev.*, vol. 75, no. 8, pp. 1208–1225, Apr. 1949, doi: 10.1103/PhysRev.75.1208.

[20] N. Aymerich et al., "*New reliability mechanisms in memory design for sub-22nm technologies,*" in *2011 IEEE 17th International On-Line Testing Symposium*, 2011, pp. 111–114, doi: 10.1109/IOLTS.2011.5993820.

[21] G. Bidal et al., "*First CMOS integration of Ultra Thin Body and BOX (UTB2) structures on bulk Direct Silicon Bonded (DSB) wafer with multi-surface orientations,*" in *Technical Digest - International Electron Devices Meeting, IEDM*, 2009, pp. 677–680, doi: 10.1109/IEDM.2009.5424247.

[22] R. Coquand, "Démonstration de l'intérêt des dispositifs multi-grilles auto-alignées pour les nœuds sub-10nm." PhD Thesis, Univ. Grenoble Alpes, 2013.

[23] C. Wee, S. Maikop, and C.-Y. Yu, "Mobility-enhancement technologies," *IEEE Circuits Devices Mag.*, vol. 21, pp. 21–36, 2005.

[24] M. Yang et al., "Hybrid-Orientation Technology (HOT): Opportunities and challenges," *IEEE Trans. Electron Dev.*, vol. 53, pp. 965–978, 2006, doi: 10.1109/TED.2006.872693.

[25] P. Packan et al., "*High performance Hi-K + metal gate strain enhanced transistors on (110) silicon,*" in *IEDM Technical Digest - International Electron Devices Meeting, IEDM*, 2008, pp. 63–66, doi: 10.1109/IEDM.2008.4796614.

[26] F. Andrieu et al., *"Impact of mobility boosters (XsSOI, CESL, TiN gate) on the performance of <100> or <110> oriented FDSOI cMOSFETs for the 32nm node,"* in *Digest of Technical Papers - Symposium on VLSI Technology*, 2007, pp. 50–51, doi: 10.1109/VLSIT.2007.4339723.

[27] G. Bidal et al., *"Planar bulk+ technology using TiN/Hf-based gate stack for low power applications,"* in *Digest of Technical Papers - Symposium on VLSI Technology*, 2008, pp. 146–147, doi: 10.1109/VLSIT.2008.4588596.

[28] I. Ben Akkez et al., *"Impact of substrate orientation on Ultra Thin BOX Fully Depleted SOI electrical performances,"* in *2012 13th International Conference on Ultimate Integration on Silicon, ULIS 2012*, 2012, pp. 177–180, doi: 10.1109/ULIS.2012.6193386.

[29] H. J. Cho et al., "Bulk planar 20nm high-k/metal gate CMOS technology platform for low power and high performance applications," 2011, doi: 10.1109/IEDM.2011.6131556.

[30] H. Shang et al., *"High performance bulk planar 20nm CMOS technology for low power mobile applications,"* in *Digest of Technical Papers - Symposium on VLSI Technology*, 2012, pp. 129–130, doi: 10.1109/VLSIT.2012.6242495.

[31] A. Khakifirooz et al., *"Fully depleted extremely thin SOI for mainstream 20nm low-power technology and beyond,"* in *Digest of Technical Papers - IEEE International Solid-State Circuits Conference*, 2010, vol. 53, pp. 152–153, doi: 10.1109/ISSCC.2010.5434014.

[32] C. Maleville, *"Extending planar device roadmap beyond node 20nm through ultra thin body technology,"* in *International Symposium on VLSI Technology, Systems, and Applications, Proceedings*, 2011, pp. 130–133, doi: 10.1109/VTSA.2011.5872261.

[33] Y. K. Choi et al., "Ultrathin-body SOI MOSFET for deep-sub-tenth micron era," *IEEE Electron Device Lett.*, vol. 21, no. 5, pp. 254–255, 2000, doi: 10.1109/55.841313.

[34] M. Jurczak et al., "Silicon-on-nothing (son)-an innovative process for advanced cmos malgorzata jurczak," *IEEE Trans. Electron Devices*, vol. 47, no. 11, pp. 2179–2187, 2000, doi: 10.1109/16.877181.

[35] N. Loubet, T. Kormann, G. Chabanne, S. Denorme, and D. Dutartre, "Selective etching of $Si_{1-x}Ge_x$ versus Si with gaseous HCl for the formation of advanced CMOS devices," *Thin Solid Films*, vol. 517, no. 1, pp. 93–97, 2008, doi: 10.1016/j.tsf.2008.08.081.

[36] S. Monfray et al., *"SON (Silicon-On-Nothing) technological CMOS platform: Highly performant devices and SRAM cells,"* in *Technical Digest - International Electron Devices Meeting, IEDM*, 2004, pp. 635–638, doi: 10.1109/iedm.2004.1419246.

[37] S. Monfray, F. Boeuf, P. Coronel, G. Bidal, S. Denorme, and T. Skotnicki, *"Silicon-On-Nothing (SON) applications for low power technologies,"* in *Proceedings - 2008 IEEE International Conference on Integrated Circuit Design and Technology, ICICDT*, 2008, pp. 1–4, doi: 10.1109/ICICDT.2008.4567232.

[38] S. Monfray et al., *"A solution for an ideal planar multi-gates process for ultimate CMOS?,"* in *2010 International Electron Devices Meeting*, 2010, pp. 11.2.1–11.2.4, doi: 10.1109/IEDM.2010.5703339.

[39] F. Balestra et al., "Double gate silicon on insulator with volume inversion: A new device with greatly enhanced performance," *IEEE Electron Dev. Lett.*, vol. EDL8, pp. 410–412, 1987, doi: 10.1109/EDL.1987.26677.

[40] T. Skotnicki et al., "Innovative materials, devices, and CMOS technologies for low-power mobile multimedia," *IEEE Trans. Electron Devices*, vol. 55, no. 1, pp. 96–130, 2008, doi: 10.1109/TED.2007.911338.

[41] J. Widiez et al., *"Experimental gate misalignment analysis on double gate SOI MOSFETs,"* in *Proceedings - IEEE International SOI Conference*, 2004, pp. 185–186, doi: 10.1109/soi.2004.1391609.

[42] J. L. Huguenin et al., *"Hybrid localized SOI/bulk technology for low power system-on-chip,"* in *Digest of Technical Papers - Symposium on VLSI Technology*, 2010, pp. 59–60, doi: 10.1109/VLSIT.2010.5556119.

[43] N. Planes et al., *"28 nm FDSOI technology platform for high-speed low-voltage digital applications,"* in *Digest of Technical Papers - Symposium on VLSI Technology*, 2012, pp. 133–134, doi: 10.1109/VLSIT.2012.6242497.

[44] C.-H. Choi et al., *"C-V and gate tunneling current characterization of ultra-thin gate oxide MOS (tox = 1.3-1.8 nm),"* in *IEEE VLSI Tech. Symp. Dig.*, 1999, pp. 63–64, doi: 10.1109/VLSIT.1999.799341.

[45] P. Packan et al., "High performance 32nm logic technology featuring 2nd generation high-k + metal gate transistors," 2009, doi: 10.1109/IEDM.2009.5424253.

[46] C. Fenouillet-Beranger et al., "Hybrid FDSOI/Bulk high-k/Metal gate platform for Low Power (LP) multimedia technology," 2009, doi: 10.1109/IEDM.2009.5424251.

[47] K. Cheng et al., "High performance extremely thin SOI (ETSOI) hybrid CMOS with Si channel NFET and strained SiGe channel PFET," 2012, doi: 10.1109/IEDM.2012.6479063.

[48] O. Weber et al., "High immunity to threshold voltage variability in undoped ultra-thin FDSOI MOSFETs and its physical understanding," 2008, doi: 10.1109/IEDM.2008.4796663.

[49] F. Andrieu et al., *"Low leakage and low variability ultra-thin body and buried oxide (UT2B) SOI Technology for 20nm Low Power MOS and Beyond,"* in *VLSI Symposium*, 2010, pp. 57–58, doi: 10.1109/VLSIT.2010.5556122.

[50] J. Mazurier et al., "On the variability in planar FDSOI technology: From MOSFETs to SRAM cells," *IEEE Trans. Electron Devices*, vol. 58, no. 8, pp. 2326–2336, 2011, doi: 10.1109/TED.2011.2157162.

[51] D. Hisamoto et al., *"A folded-channel MOSFET for deep-sub-tenth micron era,"* in *International Electron Devices Meeting 1998. Technical Digest (Cat. No.98CH36217)*, 1998, pp. 1032–1034, doi: 10.1109/IEDM.1998.746531.

[52] X. Huang et al., *"Sub 50-nm FinFET: PMOS,"* in *IEEE IEDM Tech. Dig*, 1999, pp. 67–70, doi: 10.1109/IEDM.1999.823848.

[53] D. Hisamoto, *"Critical feature size of device manufacturing for dominating MOSFET evolutions,"* in *2020 4th IEEE Electron Devices Technology Manufacturing Conference (EDTM)*, 2020, p. 1, doi: 10.1109/EDTM47692.2020.9117920.

[54] A. B. Sachid and C. Hu, "A little known benefit of FinFET over Planar MOSFET in highperformance circuits at advanced technology nodes," 2012, doi: 10.1109/SOI.2012.6404367.

[55] C. H. Lin, J. Chang, M. Guillorn, A. Bryant, P. Oldiges, and W. Haensch, *"Non-planar device architecture for 15nm node: FinFET or trigate?,"* in *2010 IEEE International SOI Conference (SOI)*, 2010, pp. 1–2, doi: 10.1109/SOI.2010.5641060.

[56] C. Auth et al., *"A 22nm high performance and low-power CMOS technology featuring fully-depleted tri-gate transistors, self-aligned contacts and high density MIM capacitors,"* in *Digest of Technical Papers - Symposium on VLSI Technology*, 2012, pp. 131–132, doi: 10.1109/VLSIT.2012.6242496.

[57] C. C. Wu et al., "High performance 22/20nm FinFET CMOS devices with advanced high-K/metal gate scheme," 2010, doi: 10.1109/IEDM.2010.5703430.

[58] J. Kavalieros et al., *"Tri-gate transistor architecture with high-k gate dielectrics, metal gates, and strain engineering,"* in *2006 Symposium on VLSI Technology, 2006. Digest of Technical Papers.*, pp. 50–51, doi: 10.1109/VLSIT.2006.1705211.

[59] K. Takahashi, A. T. Putra, K. Shimizu, and T. Hiramoto, "FinFETs with both large body factor and high drive-current," 2007, doi: 10.1109/ISDRS.2007.4422283.

[60] H. Kawasaki et al., *"FinFET process and integration technology for high performance LSI in 22 nm node and beyond,"* in *Extended Abstracts of the 7th International Workshop on Junction Technology, IWJT 2007,* 2007, pp. 3–8, doi: 10.1109/IWJT.2007.4279933.

[61] S. Inaba et al., *"FinFET: The prospective multi-gate device for future SoC applications,"* in *ESSCIRC 2006 - Proceedings of the 32nd European Solid-State Circuits Conference,* 2006, pp. 50–53, doi: 10.1109/ESSCIR.2006.307528.

[62] Y. K. Choi et al., *"Sub-20nm CMOS FinFET technologies,"* in *Technical Digest - International Electron Devices Meeting,* 2001, pp. 421–424, doi: 10.1109/iedm.2001.979526.

[63] T. Yamashita et al., *"Sub-25nm FinFET with advanced fin formation and short channel effect engineering,"* in *Symposium on VLSI Technology - Digest of Technical Papers,* 2011, pp. 14–15.

[64] A. Nainani et al., *"Is strain engineering scalable in FinFET era?: Teaching the old dog some new tricks,"* in *2012 International Electron Devices Meeting,* 2012, pp. 18.3.1–18.3.4, doi: 10.1109/IEDM.2012.6479065.

[65] A. Veloso et al., *"Capping-metal gate integration technology for multiple-VT CMOS in MuGFETs,"* in *2008 IEEE International SOI Conference,* 2008, pp. 119–120, doi: 10.1109/SOI.2008.4656323.

[66] A. Veloso et al., *"Gate-last vs. gate-first technology for aggressively scaled EOT logic/RF CMOS,"* in *2011 Symposium on VLSI Technology - Digest of Technical Papers,* 2011, pp. 34–35.

[67] B. S. Doyle et al., "High performance fully-depleted Tri-Gate CMOS transistors," *IEEE Electron Dev. Lett.,* vol. 24, no. 4, pp. 263–265, 2003, doi: 10.1109/LED.2003.810888.

[68] M. Saitoh, Y. Nakabayashi, K. Ota, K. Uchida, and T. Numata, *"Understanding of short-channel mobility in tri-gate nanowire MOSFETs and enhanced stress memorization technique for performance improvement,"* in *2010 International Electron Devices Meeting,* 2010, pp. 34.3.1–34.3.4, doi: 10.1109/IEDM.2010.5703475.

[69] M. Saitoh, Y. Nakabayashi, K. Ota, K. Uchida, and T. Numata, "Performance improvement by stress memorization technique in trigate silicon nanowire MOSFETs," *IEEE Electron Device Lett.,* vol. 33, no. 1, pp. 8–10, 2012, doi: 10.1109/LED.2011.2171315.

[70] P. Hashemi, L. Gomez, and J. L. Hoyt, "Gate-all-around n-MOSFETs with uniaxial tensile strain-induced performance enhancement scalable to sub-10-nm nanowire diameter," *IEEE Electron Device Lett.,* vol. 30, no. 4, pp. 401–403, 2009, doi: 10.1109/LED.2009.2013877.

[71] M. Saitoh, K. Ota, C. Tanaka, K. Uchida, and T. Numata, *"10nm-diameter tri-gate silicon nanowire MOSFETs with enhanced high-field transport and Vth tunability through thin BOX,"* in *2012 Symposium on VLSI Technology (VLSIT),* 2012, pp. 11–12, doi: 10.1109/VLSIT.2012.6242436.

[72] S. Barraud et al., *"Scaling of Ω-gate SOI nanowire N- and P-FET down to 10nm gate length: Size- and orientation-dependent strain effects,"* in *2013 Symposium on VLSI Technology,* 2013, pp. T230–T231.

[73] G. A. Armstrong and C. K. Maiti, *TCAD for Si, SiGe, and GaAs Integrated Circuits.* The Institution of Engineering and Technology (IET), UK, 2008.

[74] X. Zhang, D. Connelly, H. Takeuchi, M. Hytha, R. J. Mears, and T. K. Liu, "Comparison of SOI versus bulk FinFET technologies for 6T-SRAM voltage scaling at the 7-/8-nm node," *IEEE Trans. Electron Devices,* vol. 64, no. 1, pp. 329–332, 2017, doi: 10.1109/TED.2016.2626397.

[75] T. B. Hook, *"Fully depleted devices for designers: FDSOI and FinFETs,"* in *Proceedings of the IEEE 2012 Custom Integrated Circuits Conference,* 2012, pp. 1–7, doi: 10.1109/CICC.2012.6330653.

[76] T. P. Dash, S. Dey, S. Das, J. Jena, E. Mohapatra, and C. K. Maiti, "Performance comparison of strained-SiGe and bulk-Si channel FinFETs at 7 nm technology node," *J. Micromechanics Microengineering*, vol. 29, no. 10, 2019, doi: 10.1088/1361-6439/ab31c8.

[77] A. Redolfi et al., "Bulk FinFET fabrication with new approaches for oxide topography control using dry removal techniques," *Solid-State Electronics*, 2012, vol. 71, pp. 106–112, doi: 10.1016/j.sse.2011.10.029.

[78] A. Marshall, *"Designing with FinFET technology,"* in *2014 International SoC Design Conference (ISOCC)*, 2014, pp. 30–31, doi: 10.1109/ISOCC.2014.7087569.

[79] C. Auth et al., *"A 10nm high performance and low-power CMOS technology featuring 3rd generation FinFET transistors, Self-Aligned Quad Patterning, contact over active gate and cobalt local interconnects,"* in *2017 IEEE International Electron Devices Meeting (IEDM)*, 2017, pp. 29.1.1–29.1.4, doi: 10.1109/IEDM.2017.8268472.

[80] Silvaco International, VictoryProcess user manual, 2018.

[81] Silvaco International, VictoryDevice user manual, 2018.

[82] Silvaco International, ATLAS device simulator, 2018.

[83] N. Nagashima et al., *"Scalable eSiGe S/D technology with less layout dependence for 45-nm generation,"* in *2006 Symposium on VLSI Technology, 2006. Digest of Technical Papers*, 2006, pp. 64–65, doi: 10.1109/VLSIT.2006.1705218.

[84] A. Kranti and G. A. Armstrong, "Design and optimization of FinFETs for ultra-low-voltage analog applications," *IEEE Trans. Electron Devices*, vol. 54, no. 12, pp. 3308–3316, 2007, doi: 10.1109/TED.2007.908596.

[85] M. Choi, V. Moroz, L. Smith, and O. Penzin, *"14 nm FinFET stress engineering with epitaxial SiGe source/drain,"* in *2012 International Silicon-Germanium Technology and Device Meeting (ISTDM)*, 2012, pp. 1–2, doi: 10.1109/ISTDM.2012.6222469.

[86] G. Wang et al., "Integration of highly-strained SiGe materials in 14 nm and beyond nodes FinFET technology," *Solid. State. Electron.*, vol. 103, pp. 222–228, 2015, doi: 10.1016/j.sse.2014.07.008.

[87] K. Tan et al., "Strained p-channel FinFETs with extended Π-shaped silicon-germanium source and drain stressors," *IEEE Electron Device Lett.*, vol. 28, no. 10, pp. 905–908, 2007, doi: 10.1109/LED.2007.905406.

[88] N. A. F. Othman, S. F. W. M. Hatta, and N. Soin, "Impact of channel, stress-relaxed buffer, and S/D $Si_{1-x}Ge_x$ stressor on the performance of 7-nm FinFET CMOS design with the implementation of stress engineering," *J. Electron. Mater.*, vol. 47, no. 4, pp. 2337–2347, 2018, doi: 10.1007/s11664-017-6058-8.

[89] Silvaco International, VictoryStress user manual, 2018.

[90] V. V. Iyengar, A. Kottantharayil, F. M. Tranjan, M. Jurczak, and K. De Meyer, "Extraction of the top and sidewall mobility in FinFETs and the impact of fin-patterning processes and gate dielectrics on mobility," *IEEE Trans. Electron Devices*, vol. 54, no. 5, pp. 1177–1184, 2007, doi: 10.1109/TED.2007.894937.

[91] C. D. Young et al., "(110) and (100) Sidewall-oriented FinFETs: A performance and reliability investigation," *Solid-State Electronics*, 2012, vol. 78, pp. 2–10, doi: 10.1016/j.sse.2012.05.045.

[92] T. Rudenko, V. Kilchytska, N. Collaert, M. Jurczak, A. Nazarov, and D. Flandre, "Carrier mobility in undoped triple-gate FinFET structures and limitations of its description in terms of top and sidewall channel mobilities," *IEEE Trans. Electron Devices*, vol. 55, no. 12, pp. 3532–3541, 2008, doi: 10.1109/TED.2008.2006776.

[93] S. Mochizuki et al., "Quantification of local strain distributions in nanoscale strained SiGe FinFET structures," *J. Appl. Phys.*, vol. 122, no. 13, p. 135705, 2017, doi: 10.1063/1.4991472.

6 Strain-Engineered FinFETs at NanoScale

As the gate length is reduced, the difficulties in managing short channel effects in planar bulk MOSFET architectures lead to a preference for FinFET and FDSOI architectures [1]. As we have seen, bulk MOSFET technology reaches its limits, especially when electrostatic control is lost. The improvements proposed now are no longer sufficient to the design rules of sub-20 nm technologies [2], and this is why other architectures of transistors are being proposed and studied. Miniaturization takes us to the limit of the conventional transistor (sometimes called bulk), and new solutions have been proposed to enable the manufacture of performing transistors to the dimensions required by the IRDS [3] in the years to come. With very aggressive gate sizes (of the order of 20 nm), the electrostatic control of the channel becomes very critical. This is why new architectures, thin film, or even with a partially or gate-all-around can improve this control. We will describe briefly here these different technological evolutions, from bulk architecture to thin-film architectures.

Variability has become an important design issue from the 45 nm technology node. It has been reported that grain-orientation induced threshold voltage variation is the dominant source of the variations in FinFET devices. It is also worth noting that the effect of work function variation (WFV) increases with scaling. The number of grains on the gate reduces as devices scale down. WFV has been identified as a new source of random variability in metal-gate devices. In this chapter, we present an accurate physical model for WFV relevant to ultra-short channel devices. For modeling the WFV effect, there are several existing approaches, namely, the weighted average model [4], the "atomistic" simulation approach [5], and the network model [6]. In the network model, each grain on the gate is replaced with a transistor, and therefore, each device is modeled by a network of series and parallel transistors. By simulating the network (using circuit simulator) considering the boundaries conditions, the effect of WFV is considered. The main drawback of the network model is that in this model, it is assumed that the charge distribution of any specific region in the channel is determined solely by the grain above the given region.

As we have discussed in earlier chapters, the architecture of the transistors has improved over the years to evolve toward the use of new gate materials as well as the introduction of mechanical stresses (by processes or by the addition of material at the level of the transistor). Although these improvements have allowed bringing the technology on bulk silicon to the 20 nm technological node [7], these will not be sufficient to respect the specifications of the technologies to come in terms of performance (I_{ON}, I_{OFF}) and especially gate control for dimensions less than 20 nm [8]. There are two (industrially viable) solutions for reducing leakage currents of the bulk structure. The first involves the use of a thin-film structure

MOSFET transistors are fabricated on a Si layer isolated from the substrate by a layer of SiO_2 (BOX). Thanks to a very thin Si thickness (less than 10 nm), the entire channel is controlled by the gate electric field, which improves the electrostatic control. The first simulations show that the leakage current is even smaller. It has therefore been estimated that the gate length must not be greater than four times the film thickness to maintain good control over the channel. We can also add that a thin film allows the formation of steep junctions for better control of the gate length. The ITRS [9] anticipates that bulk technology will be replaced by Silicon-on-Insulator technologies, also called thin-film technologies. These architectures make it possible to better control short channel effects. The reduction of silicon thickness allows both a reduction of leakage current in the substrate and very good control of the gate on the channel.

In recent years, there has been a major structural transition from planar 2D to 3D FinFETs in Si-technology of transistors providing improvement of electrical characteristics for advanced devices [10]. A further evolution to stacked gate-all-around channels (SGAA) as nanowires (NW) [11] or nanosheets (NS) [12], is expected to overcome the limitations of FinFETs for sub-7 nm nodes. The manipulation of the elastic strains in the channel of the transistor is in all cases expected to improve device characteristics further by increasing the mobility of charge carriers [13]. As strain engineering in such devices is based on the manipulation of structure and material properties, the intricate 3D structure of SGAA renders the task particularly challenging [14]. The fabrication flow of SGAA is similar to FinFETs but is based on the formation of fin-patterned Si/ SiGe multilayers instead of a single semiconductor. Therefore, the lattice mismatch between Si and SiGe generates significant mechanical stress and interactions from the beginning of the integration process. In this framework, here we explore the unique mechanisms of SGAA transistors from a theoretical and experimental perspective.

The SiGe source/drain stressor (e- SiGe) technique has emerged as a consistent performance booster for advanced devices below 14 nm technology node whose process simulation has been discussed extensively in Chapter 5. The design and optimization of FinFETs at this nanoscale regime are extremely important [15]. The omnipresent residual stress is now becoming an important source of variability in advanced VL Si technologies that influence circuit performance. Also, in deeply scaled technologies, process and environment variations become other sources of variabilities [16, 17]. The sources of significant variability are the roughness of the lines induced by the lithography and etching and the granularity of the metal gate linked to deposition conditions. The gate may be irregular and show grains due to its polycrystalline character. The impact of these sources of variability is studied by 3D atomistic simulations.

Metal grain granularity (MGG) is an issue mainly connected with the "gate first" process technology, where the gate metal is deposited before any high-temperature annealing procedure [18]. During high-temperature processing, the nominally amorphous metal gate material becomes polycrystalline. This results in the formation of grains with different crystallographic orientations and differing metal work functions. Within the case of polysilicon gates, the interfaces between grains cause Fermi level pinning and doping non-uniformity as a consequence of rapid diffusion along grain boundaries [19]. MGG induced random WFV is another important source of random variability for the technology nodes that use a high-k/metal gate stack [20]. Depending on their crystal orientations, the work function of each metal grain varies randomly. Therefore, instead of a single work function value, the transistor gate contains multiple grains with different work function values that cause random fluctuation of device performance parameters [21, 22].

MGG and random discrete dopants (RDDs) are major sources of process-induced variability in FinFETs [23, 24]. The nature of the metal grain variability is the domains, or grains, that arise in the metal during the contact formation with the gate. Among other technologies that have been developed to increase the capacitance of the gate contact, one uses the gate dielectric with high-k and metal gates. The domains with different orientations arise in the metallic contacts, have random forms and orientations, depending on the deposited material, and also present different values of work function, which has a detrimental effect on the device performance. In the case of MGG, the parameters involved are the average size of the grains, which is controlled through the number of nucleation points, the possible orientations, their probabilities, and the work function that each orientation has [25].

To model this source of variability, one of the options is to split the gate as if it were composed of several gates in parallel and apply the model to take into account the effect of this gate partitioning. Another approach is to model the gate with square grains that cover the area of the gate and apply to each of these grains a different value of work function, and then simulate the device. These squares can have different sizes and orientations depending on the gate material. Another variability issue, such as RDD that becomes important when approaching nanoscale devices [26, 27] is the stochastic nature of individual discrete dopants, which is considered using RDD simulations. The influence of RDD on the device properties is considered next. As a consequence, variability sources such as RDD and MGG affect the threshold voltage (V_T), drain-induced barrier lowering (DIBL), and subthreshold slope (SS) in p-channel FinFETs. The electrical characteristics variation has been analyzed by calculating the mean and standard deviation of these devices.

6.1   DESIGN AND SIMULATION AT 7N

In this chapter, we study the stress-induced variability in aggressively scaled (at 7N) Si-channel FinFETs with epitaxially grown SiGe (e- SiGe) stressors in the source and drain regions. The stress distribution analysis is performed with the help of TCAD mechanical stress simulations. We have generated the stress maps in FinFETs with various scaled dimensions of fin length, height, and width. The effects of residual strain/stress on variability due to metal grain granularity and RDD in strain-engineered FinFETs (with 50 configurations) have been presented. Furthermore, the device critical parameters such as threshold voltage, I_{on}, I_{off}, subthreshold slope variation due to RDD, and MGG are examined. Finally, we calculate the mean and standard deviation of these parameters (QQ plots) to quantify the variability. We use an extensive 3D TCAD simulation framework [28] to demonstrate how the elastic stress models can be applied to the simulation of stress transfer using source/drain SiGe -epi layer in 7 nm technology nodes for p-FinFETs.

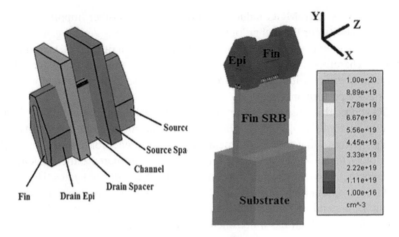

FIGURE 6.1   Detail of the fin structure in FinFET at 7N technology node.

Figure 6.1 shows the details of the structure considered in simulation showing the dumbbell-shaped SiGe source/drain stressor [29] and the doping in the device. Stress is generated due to the lattice mismatch, thermal mismatch, and intrinsic stress, which has been taken here between Si and SiGe layers. All the technological parameters are listed in Table 6.1.

Stress and strain in the lattice modify the material properties. For the advanced technology nodes, it becomes increasingly difficult for empirical mobility modeling to keep up with shrinking device dimensions and new technological approaches. Classical device simulation based on empirical mobility models also suffers from poor accuracy and predictive power as an entirely empirical, geometry-dependent

TABLE 6.1

Technology and Geometry Parameters are Considered in the Simulation

Design Parameters	7 nm Node	Units
Fin Height	30	nm
Fin Width	5	nm
Fin SRB Thickness	50	nm
Oxide Thickness	0.5	nm
High-k (HfO$_2$)Thickness	1.5	nm
Gate Length	14	nm
Spacer Length	7	nm
Epi length	14	nm
Epi Top Width	8	nm
Epi Bottom Width	14	nm
Epi Middle Width	25	nm
Well Doping	1e20	cm^{-3}
Epi Doping	2e20	cm^{-3}
Substrate Doping	1e16	cm^{-3}

TABLE 6.2

Average Stress Values and Piezoresistivity Factors are Considered in the Simulation

Material	n-type		p-type	
10^{-12} cm²/dyne	<100>	<110>	<100>	<110>
π_{11}	−102.2	−31.1	6.6	71.8
π_{12}	53.4	−17.5	−1.1	−66.3
π_{13}	53.4	53.4	−1.1	−1.1

mobility needs to be used to maintain the required accuracy [30]. Physical device modeling allows for more accurate and robust predictions of device performance and allows us to assess novel process options found in 7 nm and 5 nm technology nodes [31]. Piezoresistivity constants [32] for different wafer orientations are used for mobility enhancement evaluation in simulation. Mobility enhancement factors are calculated by the use of average stress values and piezoresistivity factors as shown in Table 6.2.

$$\frac{\Delta\mu_{xx}^{DG}}{\mu} = \left(+\pi_{11}\sigma_{xx}\right)\times\left(1+\pi_{12}\sigma_{yy}\right)\times\left(1+\pi_{13}\sigma_{zz}\right)-1 \tag{6.1}$$

$$\frac{\Delta\mu_{xx}^{TG}}{\mu} = \frac{2}{3}\frac{\Delta\mu_{xx}^{DG}}{\mu} + \frac{1}{3}\left[\left(1+\pi_{11}\sigma_{xx}\right)\times\left(1+\pi_{12}\sigma_{yy}\right)\times\left(1+\pi_{13}\sigma_{zz}\right)-1\right] \tag{6.2}$$

As the semiconductors are piezoresistive materials, their resistivity changes as stress or strain are applied. Toward the prediction of the strain-induced mobility enhancements in nanodevices, the bulk piezoresistance model is commonly used. The physical basis of mobility enhancement in the inversion layer of biaxially tensile strained- Si MOSFETs is well understood. However, the charge transport effects of uniaxial strain are not fully understood from a theoretical standpoint. Therefore, to predict the expected strain-induced mobility enhancements, bulk piezoresistance theory is used in this work [33]. Device simulations have been performed for FinFET structures using the physical-model-based VSP and MINIMOS tools [34, 35] which allow the combination of different carrier models within a self-consistent loop, for example, classically calculated carriers in the contact regions and quantum-mechanically confined carriers in the channel region [36–38].

The mobility depends on the orientation of the sidewall and the channel. For simulation, the sidewall and channel orientation were taken as (110) and (100), respectively, as the orientation of the Si surface is an important parameter for thermal oxidation of Si. In this work, the simulation is performed using the 3D Drift-Diffusion [39, 40] and Density-Gradient [41] model. A coupled-mode space representation is used [42, 43]. The DD/DG formulations have been solved self-consistently with the 3D Poisson equation. The raised S/D resistance and leakage across the SRB are covered by the Drift-Diffusion/density-gradient simulation. The source to drain tunneling is described by this formalism, which is extremely necessary for gate lengths below 10 nm. In a semi-classical formalism, the local potential is still not well understood. The local potential is generated by the impurities and that impurity potential depends on the electron wavelength. In RDD simulation, the discrete dopants are distributed in the source/drain region of the device randomly. The small dimensions of the FinFET lead to the use of the DG model which takes quantum transport effects into account.

In the classical model, the current relation for FinFETs is given by

$$I_D = I_{D,top} + I_{D,lateral} \tag{6.3}$$

The drain current in the FinFETs depends on the distribution of electrons as well as the electric field. The current in FinFETs also depends on the carrier mobility which is given by

$$\mu_{eff}(W) = \mu_{lateral} + (\mu_{top} - \mu_{lateral})\frac{W}{2H + W} \tag{6.4}$$

where $\mu_{lateral}$ is the mobility of channel in the lateral side of the fin, μ_{top} is the mobility of channel on the top side of the fin, μ_{eff} is the effective mobility in the FinFET, W is the fin width and H is the fin height.

Further, the threshold voltage can be expressed as

$$V_{th} = V_{FB} + 2\varnothing_F + K_B\sqrt{2\varnothing_F - V_{FB}} \tag{6.5}$$

where V_{FB} is flat band voltage, $\varnothing_F$ is the Fermi potential and $K_B = \dfrac{\sqrt{2q\varepsilon_{0x}N_B}}{C_{ox}}$, is the body coefficient. (q is the electronic charge, C_{Ox} is gate oxide capacitance, ε_{0x} is gate oxide relative permittivity, N_B substrate doping). Equation (6.5) has been used to extract threshold voltage which has been shown in the next section.

6.2   DESIGN ISSUES

The main focus of the work is the performance evaluation of 14 nm long channel FinFET for three different sources of variability, viz., (a) geometry variation (Length (L), Height (H), and Width (W)), (b) metal grain granularity, and (c) RDD. In the first case (a), as geometry variation leads to a change in stress developed in the fin, we have also shown the stress variation in support of electrical performance enhancement. In the latter two cases (b, c), the stress variation has not been shown explicitly, rather the change in electrical performance considering MGG and RDD effect into account has been shown.

6.2.1   EFFECTS OF GE CONTENT VARIATION

The longitudinal stress obtained from simulation has been mapped in 2D and has been shown. Figure 6.2 illustrates the longitudinal stress (S_{zz}) variation in the different regions for the gate length (7 nm) with 10% Ge-content in SiGe (source/drain

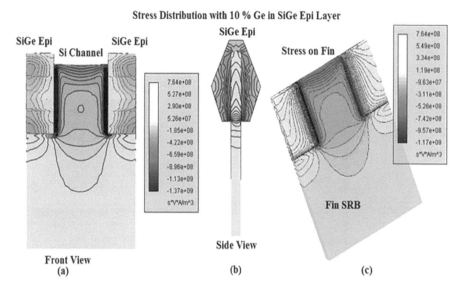

FIGURE 6.2   Stress distribution with Ge 10% in the SiGe Epi layer.

stressor). A large compressive lateral and vertical tensile stress is observed in the silicon channel confined by the two embedded SiGe stressors. For quantitative analysis of the distribution of stress in the longitudinal and perpendicular direction the stress component, S_{zz} and S_{yy} are plotted along the channel direction as shown in Figures 6.3a and 6.3b, respectively. The stress variation due to change in Ge content (10–50%) is shown. The peak S_{zz} value can be noted to be −2 GPa in the channel and +4 GPa in Source/Drain whereas the S_{yy} is insignificant in the channel region and maximum +2.42 GPa in the S/D region.

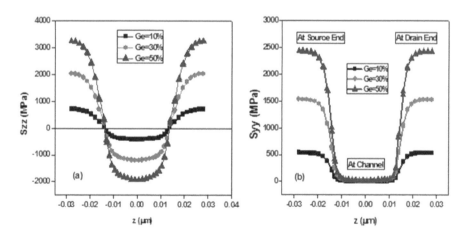

FIGURE 6.3   Variation of (a) longitudinal stress (S_{zz}), (b) perpendicular stress (S_{yy}) along the channel (from source to drain).

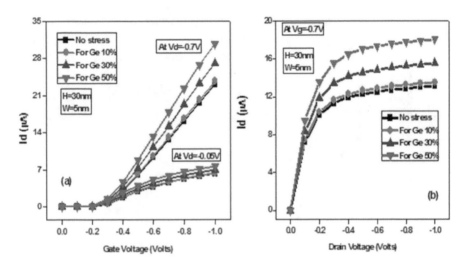

FIGURE 6.4 (a) Transfer $(I_d\text{-}V_g)$ characteristics (b) Output $(I_d\text{-}V_d)$ characteristics with various Ge contents in the SiGe Epi layer.

As with the introduction of a higher concentration of Ge, the lattice mismatch between the epitaxial SiGe layer and underneath the Si layer increases. It increases the stain amount following Equations (6.6, 6.7). The strain (ε) in the Si channel is determined by the Ge mole fraction of the adjacent $Si_{1-x}Ge_x$ source/drain region, and this is directly proportional to the volume under stress and is calculated as [44]:

$$\varepsilon = (0.0425\,x)\,x \text{ is Ge Mole fraction} \tag{6.6}$$

$$\varepsilon = \frac{L_{SiGe} - L_{Si}}{L_{Si}} \tag{6.7}$$

With a higher Ge mole fraction, silicon lattice becomes more stretched to combine with Ge lattice and forms the compound of $Si_{1-x}Ge_x$. So the longitudinal compressive stress on the channel and perpendicular tensile stress in the source and drain region increases. It leads to improvement in an overall improvement in electrical performance. The device transfer $I_d\text{-}V_g$ and the output characteristics $(I_d\text{-}V_d)$ are shown in Figure 6.4a and b, respectively. A clear enhancement in drain current is observed as strain in the fin increases as Ge content increases from 10% to 50%. It shows almost 40% improvement in drain current with 50% Ge as compared to the device with no stress.

6.3 VARIABILITY DUE TO GEOMETRY CHANGE

As fin geometry plays an important role in determining the electrical performance in non-planner structures like FinFETs, it is highly essential to study the performance variation introduced due to geometry variation.

6.3.1 GATE LENGTH VARIATION

The device has been scaled down from 35 nm to 7 nm gate length. The stress developed in the channel region has been extracted for all the devices. The stress should be compressive which has been transferred from the raised source–drain (with e- SiGe material) to the Si channel. As SiGe has a larger lattice constant compared to Si, the strain developed in Si is compressive and expressed as a negative value. The stress value obtained from simulation has been shown in Figure 6.5a in terms of a box chart. It shows a range of longitudinal stress (S_{ZZ}) values in the channel for the different sized channels of the p-type FinFET device at 25% Ge content. The stress value extracted from the device ranges from −672 MPa to −713 MPa in the 7 nm channel device with an average stress value of −693 MPa. Similarly, the stress values for other channel lengths (14 nm to 35 nm) have been shown, which shows gradual decrement with an increase in gate length. The average stress values are found to be −478 MPa, −432 MPa, −219 Mpa, and −145 MPa, respectively, for 14 nm, 21 nm, 28 nm, and 35 nm, respectively. Figure 6.5b shows the transfer characteristics (both in log and linear plot) for the above devices comparing with bulk- Si channel devices at V_d = −0.7 V. With downscaling, the drain current is expected to be higher. The enhancement in drain currents in the stress enhanced device is found to be 28.7%, 33.34%, 29.1%, 25.6%, and 22.5% for 7 nm to 35 nm long channel devices, respectively.

Further, the threshold voltages are also analyzed which have been extracted from transfer characteristics. Figure 6.6a shows the change in threshold voltage with a change in stress due to the change in Ge concentration at different gate lengths. V_T normally reduces with downscaling which becomes more aggressive with the introduction of stress. The threshold voltage obtained for the 35 nm device is −221 mV

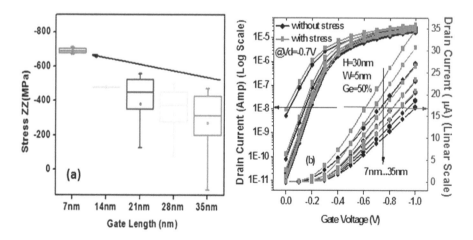

FIGURE 6.5 (a) Comparison of stress at mid-point of the channel at Ge = 25% in SiGe epi layer, devices with different gate lengths (b) I_d-V_g characteristics comparison of the devices with different gate lengths.

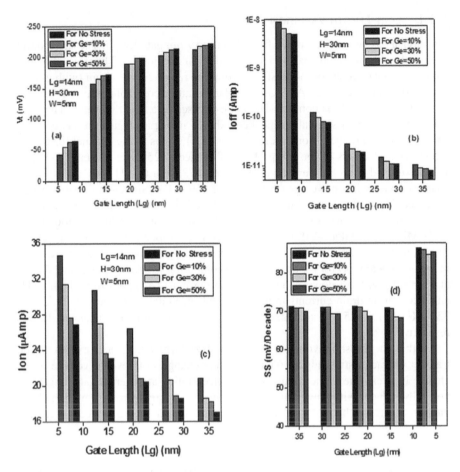

FIGURE 6.6 Comparison of (a) threshold voltage, (b) I_{off}, (c) I_{on}, (d) subthreshold slope for different strain levels (Ge percentage) for different gate length devices.

which reduces to −213 mV for 50% Ge concentration. For a 7 nm device, it is 65 mV without stress which reduces to 43 mV for 50% Ge. The off-state current (I_{off}) slightly increases with an increase in stress. With a decrease in gate length from 35 nm to 7 nm the off-state current increases in the order of 2 and it maintains the 10^{-8} Amp current for the 7 nm device shown in Figure 6.6b. Figure 6.6c shows the I_{on} values at different gate lengths and for different Ge. As mobility increases with an increase in stress due to an increase in Ge content, I_{on} is expected to be higher with an increase in Ge content. The subthreshold slope has also been shown with Ge content variation for different gate lengths. It can be observed from the simulation that an increase in Ge content shows a very small decrement in the subthreshold slope. The subthreshold slope almost remains in the range of 65–70 mV/decade for all gate lengths except marginal increment in 7 nm device that is 85 mV/decade shown in Figure 6.6d, within the range where the device can be considered to be reliable.

6.3.2 FIN HEIGHT VARIATION

Figure 6.7a shows the change in stress due to the change in fin height. As height changes, the area of contact of the fin with raised-source drain SiGe increases. This allows to e- SiGe Source/Drain to transfer more stress to the fin. It can be seen in Figure 6.7a that the stress increases with an increase in fin height. This also reflects in the drain current which has been shown in Figure 6.7b. It shows transfer characteristics for the stress enhanced device for three different fin heights compared with convention devices (no stress). This enhancement in drain currents is found to be 11.76%, 9.26%, and 8.3%, respectively, for 10 nm, 20 nm, 30 nm fin height.

Figure 6.8a shows the variation of threshold voltage for different heights with and without stress conditions. The threshold voltage reduces with an increase in fin height in both stress and no-stress condition. The V_T is found to -405.6 mV, -202.53 mV, and -167.4 mV for stress enhance devices at 10 nm, 20 nm, 30 nm height,

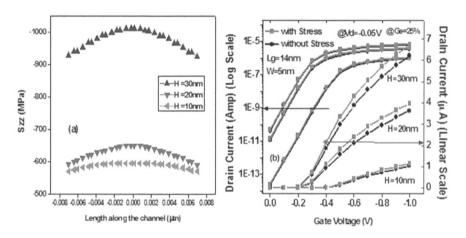

FIGURE 6.7 (a) Variation of longitudinal stress (S_{zz}) along the channel (b) I_d-V_g comparison of stress enhanced devices with the conventional device, at different fin heights.

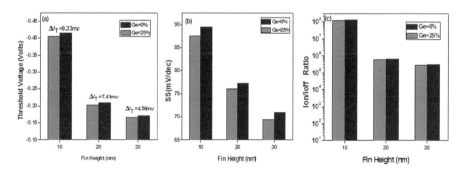

FIGURE 6.8 Comparison of (a) V_T, (b) SS, and (c) I_{on}/I_{off} ratio variation with change in fin height for with stressor and without stressors.

respectively. The change in threshold voltage in stress enhanced devices compared to no stress condition has been calculated and is shown in Figure 6.8a. The subthreshold slope also increases with a decrease in fin height. The subthreshold slope also gets higher for smaller fin height. For 50 nm and 40 nm fin height, the threshold slope remains within 70 mV/decade moreover the introduction of stress rarely affects this fin height as shown in Figure 6.8b. However, when the fin height goes below 30 nm the effect of stress on subthreshold characteristics becomes prominent. It can be observed that the lowering in subthreshold slope in Ge 25% compared to Si for devices with 30 nm, 20 nm, and 10 nm fin height. Figure 6.8c shows the I_{on}/I_{off} ratio for different fin heights for both stressed and unstressed devices. With a decrease in fin height I_{on}/I_{off} ratio remains the same at 10^5–10^6 except for 7 nm in which the ratio is found to be 10^8.

6.3.3   FIN WIDTH VARIATION

The stress in the fin also gets changed with a change in fin width. It has been found that the stress increases with a reduction in fin width or vice versa. As fin is of Si material and S/D is of SiGe material, the increment in fin width leads to stress relaxation in the fin keeping the SiGe layer thickness fixed. The 1D stress profile across the channel has been shown in Figure 6.9a. The enhancement in current due to stress has been shown in Figure 6.9b along with the width variation. The width has been varied from 3 to 10 nm keeping the length and the height fixed. The enhancement due to stress is around 6%–13% (from 10 to 3 nm) compared to a conventional device.

Figure 6.10a shows the change in threshold voltage in stress enhanced device compared to a conventional device at different fin widths. With an increase in fin width, threshold voltage reduces, which becomes more aggressive in stress enhanced devices. The threshold voltage of the stress enhanced device is 15.34 mV lower compared to conventional Si devices at 3 nm width which becomes 57.20 mV at 10 nm fin width. Figure 6.10b shows the subthreshold slope increases with the widening of

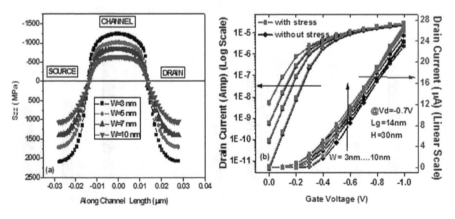

FIGURE 6.9   (a) Longitudinal stress (S_{zz}) variation along the length of the channel for different fin widths (b) I_d-V_g for different fin width.

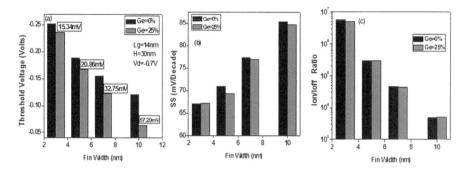

FIGURE 6.10 Comparison of (a) V_T, (b) SS, and (c) I_{on}/I_{off} ratio for different fin widths with and without stress.

fins whereas it decreases with stress compared to the no-stress condition. It seems the ON-state current does not affect much by width whereas the OFF-state current increases with an increase in width. However, the I_{on}/I_{off} ratio decreases with the widening of fins as shown in Figure 6.10c. The subthreshold slope and I_{on}/I_{off} ratio remain unaffected by the introduction of stress.

6.4 VARIABILITY DUE TO METAL GRAIN GRANULARITY

Transistors are produced using a large number of manufacturing processes. The key device parameters of a transistor are the threshold voltage, drain current and its slope below the threshold, and DIBL. However, a great complexity is involved in the production of transistors in advanced technological nodes leading to fluctuations of the electrical parameters of the transistors which are known as variability. In the following, static variability sources due to manufacturing processes, such as RDD and MGG are considered. The fluctuations induced by RDDs and metal grain roughness are considered separately.

Assume the gate length and width of the metal gate are L and W, respectively. The calculation of the number of grains on the gate surface (N_{eff}) is based on the binomial distribution model. Therefore, N_{eff} distributed on the metal-gate surface is written as follows:

$$N_{eff} = \frac{Area}{\pi \times \left(\dfrac{GS}{2}\right)^2} \tag{6.8}$$

The grain probability is mapped with work functions for easy distribution of grains with a condition that the total probability should be one. The work function of the gate metal will be a probabilistic approach rather than a deterministic value. Assuming $X_1, X_2, X_3 \ldots\ldots X_n$ to be the random variables (where $N_{eff} = X_1 + X_2 + X_3 \ldots \ldots + X_n$) that denote the number of grains having WF values of $\phi_1, \phi_2, \phi_3 \ldots\ldots \phi_n$, respectively. The probabilistic distribution of work function in a given metal can be realized as [22]

$$\phi_{Metal} = \left(\frac{X_1}{N_{eff}}\right) \cdot \phi_1 + \left(\frac{X_2}{N_{eff}}\right) \cdot \phi_2 + \left(\frac{X_3}{N_{eff}}\right) \cdot \phi_3 + \dots\dots + \left(\frac{X_n}{N_{eff}}\right) \phi_n \qquad (6.9)$$

The first term $\dfrac{X_1}{N_{eff}}$ represents the percentage of the gate area covered with grains having WF of ϕ_1. The probability of getting the above-mentioned work function can be calculated from the probability density function (PDF),$(P_X(k))$ as:

$$P_X(k) = \binom{N_{eff}}{k} \times P_1^k \times (1 - P_1)^{N_{eff} - k} \qquad (6.10)$$

In this case, two random variables X_1 and X_2 (where $N_{eff} = X_1 + X_2$) that denote the number of grains having WF values of ϕ_1 and ϕ_2 in the binomial distribution model with probabilities of P_1 and P_2 (assume $\phi_1 = 4.6$ eV and $\phi_2 = 4.35$ eV, $P_1 = 0.6$ and $P_2 = 1 - P_1 = 0.4$) respectively.

These two random variables are not independent ($X_2 = N_{eff} - X_1$). The probability of getting exactly $X_1 = k$ of such grains can be calculated by a binomial distribution where the PDF is given by:

$$P_{X_1}(k) = \frac{N_{eff}!}{k!(N_{eff} - k)!} 0.6^k 0.4^{N_{eff} - k} \qquad (6.11)$$

where $\dbinom{N_{eff}}{k} = \dfrac{N_{eff}!}{k!(N_{eff} - k)!}$ and

$$\phi_{Metal} = \left(\frac{X_1}{N_{eff}}\right)(4.6\,\mathrm{eV}) + \left(\frac{X_2}{N_{eff}}\right)(4.35\,\mathrm{eV}) = \left(\frac{X_1}{N_{eff}}\right)(4.6\,\mathrm{eV}) + \left(\frac{N_{eff} - X_1}{N_{eff}}\right)(4.35\,\mathrm{eV})$$

We consider the work function and size of the individual grains in determining the local band structure. The proposed model is much more accurate than other models [45]. Additionally, using this new model, the WFV effect can be captured using 20 device simulations, resulting in a significantly lower simulation time. It is experimentally verified that different grain orientations have a different probability of occurrence. In other words, each metal has a preferred orientation" for which the crystal structure of the metal is more stable (orientation with the highest probability of occurrence). For very high processing temperatures, all metal grains tend to grow in the preferred direction. However, for lower temperature ranges, such as those used in the fabrication of integrated circuits, metal gates of transistors consist of both the preferred and other crystal orientations. To consider the effect of adjacent metal grains in different parts of the channel, we assign an effective WF to each grain. The effect of the neighboring grains also depends on the average grain size. If the grain size is large (small), the neighboring grains have a lower (higher) effect on the charge below a given grain.

The variability can be reduced by introducing the high-k/metal gate at 45 nm [20]. In ultra-small devices, the channel is near intrinsic, therefore, the impact of random dopant fluctuation (RDF) on the threshold voltage variation is significantly reduced. It is thus important to compare the effect of other sources of variation including fluctuation in gate length, fin thickness, oxide thickness, and WFV in metal-gate FinFETs. The standard deviation of threshold voltage varies inversely with the effective electrical thickness of the gate capacitance. The polycrystalline nature of the metal gate has become a new source of statistical variability [46]. Different metal grain orientation leads to variation in threshold voltage. The statistical variation in threshold voltage has been investigated by several research groups.

Nawaz et al. [47] assumed that the grains are identically sized squares with equal weight in their contribution to an average work function, thus neglecting the effects of grain size and boundary orientation variations. When scaling device dimensions down to the deep submicron range, a statistical approach to the simulations is required to account for the various microscopically different random charge distributions in macroscopically identical devices. This approach implies an estimation of the basic design parameters (such as threshold voltage, subthreshold slope, transconductance, and driving current) averaged over the statistical ensemble of microscopically different devices, rather than predicting the characteristics of a single device with continuous doping. For the charge of each dopant to be spatially resolved in full-scale 3D atomistic simulations with fine grain, discretization is required [48]. To consider the WFV effect, a large real granular image is used as a template from which a small random section with the size of the gate is imported into the simulator. Since the "atomistic" simulation method considers the randomness of the shape of the grains, simulation conditions mimic the actual grain combinations in the gate of the metal-gate devices. Therefore, the "atomistic" simulation methodology accurately predicts the WFV effects on the device characteristics, however, this method is very time consuming and cannot be employed for circuit simulations.

MGG due to grain orientation leads to variation in the effective work function of high-k metal gate stack in nanoscale devices. The spatial variation of the work function is taken into account to model the MGG. The atomistic simulations have been performed to know the impact of metal grain orientation on the gate work function. In the general case, firstly, the gate is divided into different segments and the WF of each grain is assigned randomly based on the probabilities of different grain orientations. THE effective WF of each grain is determined based on the WF of the grain itself and the WF of the neighboring grains as well as the average grain sizes. In the simulation, we use the Poisson–Voronoi model to simulate the polycrystalline grain structure of materials. The Voronoi diagram has been used to investigate metal grains in solid-state physics [49, 50]. The MGG induced variations due to different surface orientations of metal grains lead to variability in the effective work function as well. Using the Voronoi approach, we can generate various metal-gate grain patterns, the grain depends on various factors, for example, grain position, shape, size, and orientation. For constructing the Voronoi diagram, we place the grain seeds randomly and the number of grain seeds is determined by the size of the grain. With the assumption that the grain growth rate of each grain is the same for describing metal grain, the grain boundary will locate at the perpendicular bisector of neighboring grain seeds in the

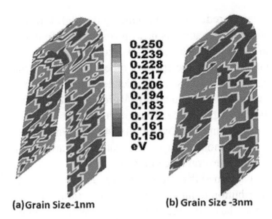

FIGURE 6.11   Work function variation in the metal gate (a) 1 nm grain size, (b) 3 nm grain size. The bar in middle represents the work function difference between metal and substrate in eV.

Voronoi diagram. For the present study of variability, we have generated two different work function patterns with two average grain sizes of 1 nm and 3 nm, respectively.

The work function distribution due to metal grain granularity (i.e., different metal grain size has been introduced which causes the so-called metal grain granularity) is presented in Figure 6.11. Figure 6.12 shows the I_d-V_g characteristics, including metal grain granularity with an average grain size of 1 nm and 3 nm, respectively, for 50 different configurations. The drain current fluctuation increases with an increase in

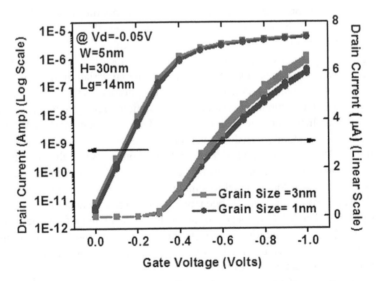

FIGURE 6.12 I_d-V_g plot of 14 nm gate length p-channel FinFETs, including metal grain granularity with an average grain size of 1 nm and 3 nm, respectively, for 50 configurations. Different metal grain orientation has been introduced which causes the so-called metal grain granularity.

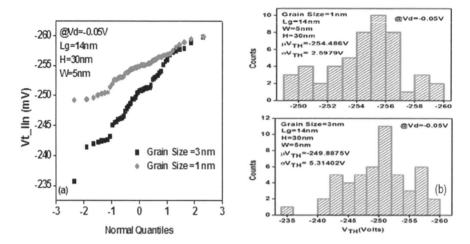

FIGURE 6.13 (a) The Q-Q test on V_T distribution due to MGG at linear drain bias. The distribution of V_T deviates more from the normal distribution on the upper tail and lower tail in both the grain sizes. (b). Histogram plots of V_T for 50 different configurations subjected to MGG as a source of statistical variability with two different grain sizes (1 nm upper, 3 nm-lower). σ stands for standard deviation and μ stands for the mean.

grain size. As the distribution of the metal grain depends on its size and orientation, the work function fluctuation slightly high for lower grain size (shown in Figure 6.11). This contributes to the potential fluctuation in the channel which leads to drain current fluctuations as shown in Figure 6.12.

Figures 6.13a and 6.13b show the statistical variation of threshold voltage for two types of grain size. The fluctuation of V_T (normal QQ plots) in the linear region is shown in Figure 6.13a. The distribution of V_T follows normal distribution except for a slight deviation on the upper tail and lower tail for both the grain sizes. For further understanding of the effect of MGG on V_T, the distribution of V_T for 50 different configurations of simulations is shown in Figure 6.13b. The mean value of V_T is found to be −254.49 mV, 249.89 mV, and the standard deviation is 2.58 mV, 5.314 mV for 1 nm, and 3 nm grain size, respectively.

From Figure 6.14a, the mean and standard deviation of OFF-state current are found to be 5.011 and 0.501 pA, respectively, considering the MGG effect with an average 1 nm grain size. The mean and standard deviation are found to be 5.919 pA and 0.0932 pA, respectively, for 3 nm grain size shown in Figure 6.14b. From Figure 6.14c, d, it is observed that the mean and standard deviation of ON current is found to be 6.011 μA and 0.039 μA, respectively, with MGG (for 1 nm grain) and 6.5367 μA and 0.0689 μA, respectively (3 nm grain). The variation in on-current is more pronounced with an increase in grain size. The statistical distribution of the sub-threshold slope due to MGG has been shown in Figure 6.14d, f. It has been compared to two different grain sizes. The mean and standard deviation are found to be 71.903 mV/dec and 0.496 mV/dec, respectively, in MGG (1 nm grain). The mean and standard deviation are found to be 72.06 mV/dec and 0.866 mV/dec, respectively, with MGG (3 nm grain).

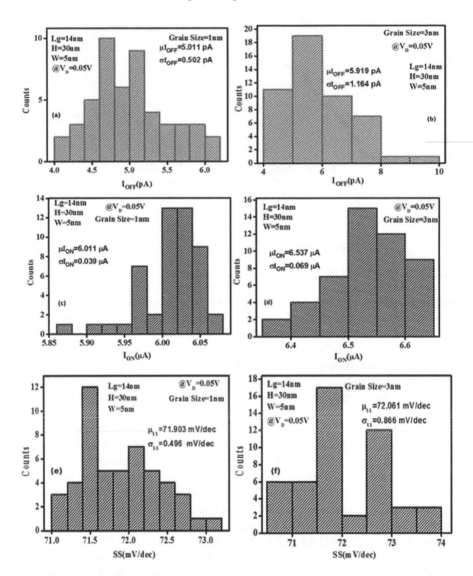

FIGURE 6.14   Histogram plots of I_{off} (a, b), I_{on} (c, d), and SS (e, f) for 50 different configurations subjected to MGG as a source of statistical variability with two different grain sizes (1 nm left, 3 nm-right), respectively. σ stands for standard deviation and μ stands for the mean.

6.5   VARIABILITY DUE TO RANDOM DISCRETE DOPANTS

The consideration of discrete dopants instead of continuous doping densities can be relevant for the modeling of small device structures. Using this method also variability simulations can be performed by randomly placing the dopants in the simulation domain as shown in Figure 6.15.

In the simulation, discrete dopants are generated randomly distributed in the simulation domain or a given segment and a list of dopants with their distinct

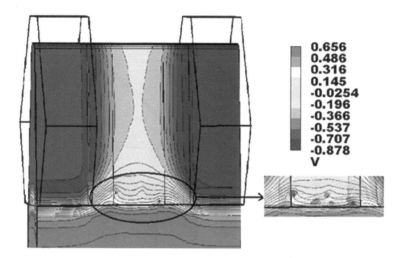

FIGURE 6.15 Illustration of (a) potential variation in the active region of the device in presence of random discrete dopant and (b) selected region of channel showing the presence of dopant. The right side bar shows the potential in volts.

positions in the device is created. The randomly generated dopants follow the continuous doping concentration as a mean value. For each grid point, the number of dopants is determined using a Poisson distribution. The position within the control volume of each grid point follows a uniform distribution. When the location and number of dopants are varied, the transfer characteristics for the devices become different. The modification of effective channel length and the height of the source/ drain barrier by the discrete dopants is demonstrated in this example which leads to a reduction of the gate control over the charge in the channel. At high drain bias, random dopant effects become more prominent. The statistical distribution of V_T due to RDF is found to follow a normal distribution. To study the impact of s, the number of dopant atoms and the positions of the given dopant atoms in the S/D region are taken into consideration. The selection of the number of dopants in each device can be determined from Poisson's distribution of the average number of dopants. The distribution of the V_T can be mathematically evaluated through the discrete convolution of the Poisson distribution with the mean value of a statistically significant region (i.e., S/D extension region). The deviation in the transfer characteristics is due to the variation in location and number of dopants. The equivalent charge density corresponding to the number of dopants suggested by Sano et al. [51] is given as:

$$n(r) = N_f \cdot \frac{k_c^3}{2\pi^2} \frac{\sin(k_c r) - (k_c r)\cos(k_c r)}{(k_c r)^3} \qquad (6.12)$$

where N_f is the normalization factor, k_c is the inverse of screening length, and r is the distance from the discrete dopant. The screening length can be calculated by using

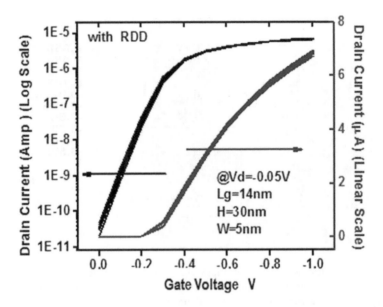

FIGURE 6.16 I_d-V_g plot of 14 nm gate length p-channel FinFETs including random discrete dopants in the channel.

the Conwell–Weisskopf model [52, 53] and defined as $k_c \approx 2(N_{D/A})^{1/3}$ where $N_{D/A}$ represent donor/acceptor concentration.

The standard deviation of V_T distribution due to RDF is modeled as:

$$\sigma_{V_{th}} = \sqrt[4]{4q^3\varepsilon_{Si}N_a\varphi_F}\,\frac{T_{ox}}{\varepsilon_{ox}}\frac{1}{\sqrt{2W*L}} \tag{6.13}$$

Variation occurs due to the presence of discrete dopants in the channel. As the channel is undoped (i.e., less doped in comparison to the S/D region), the dopants from the source–drain region may diffuse toward the channel. Their random position in the channel leads to the variation in the electrostatic of the device. The distribution of potential due to discrete dopants in the channel is shown in Figure 6.15. The variation of drain current due to RDD in the channel is shown in Figure 6.16.

Figure 6.17a shows the statistical variation of V_T for 50 different configurations of simulation conditions. It shows the mean V_T value is −195.832 mV with a standard deviation of 3.066 mV. Figure 6.17b shows the statistical distribution of the subthreshold slope of stress enhanced device due to RDD. It can be seen that the mean value of the subthreshold slope in RDD is 75.593 mV/dec and the standard deviation is 0.31306 mV/dec. Similarly, the variation in ON and OFF state current has been shown in Figure 6.17c and d. The mean value of the OFF-state current is found to be 35.9 pA with a standard deviation of 3.43 pA due to RDD. Similarly, the mean value and standard deviation of I_{on} value are obtained at 6.877 μA and 0.045 μA.

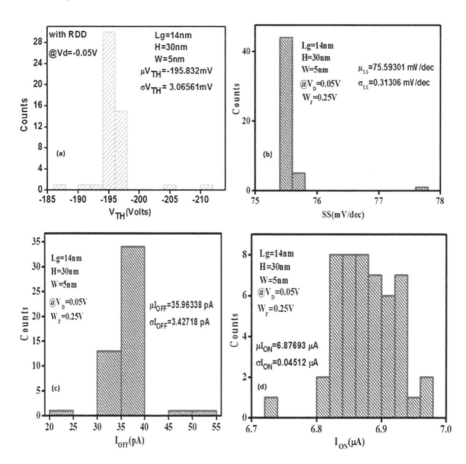

FIGURE 6.17 Histogram plots of (a) V_T, (b) SS, (c) I_{off}, and (d) I_{on} for 50 different configurations subjected to RDD as a source of statistical variability: σ stands for standard deviation and μ stands for the mean.

6.5.1 SUMMARY

We have demonstrated the role of the stress/strain on device variability in a state-of-the-art trigate FinFET with SiGe source/drain stressors at a 7 nm technology node using full 3D TCAD simulations. A comprehensive study of the physical effects that take place due to SiGe source/drain stressor in p-FinFETs has been performed. The stress mapping technique is used to quantify strain distribution in the devices. It is shown that a shallower and closer-to-channel e- SiGe stressor with higher Ge content could increase compressive stress and therefore improve the drive current through hole mobility enhancement. The possibility of tuning the stress by changing the geometry of the SiGe stressor during epitaxial growth is also demonstrated. It is seen that with the increase in fin height and fin angle I_{on} and I_{off} increase. Similarly, when the fin width increases, the variation of threshold voltage, on-state current reduces. A reduction in the fin dimension leads to a higher threshold voltage. So there exists a tradeoff between variation in current and threshold voltage with fin width scaling. Fin height control and recess of shallow trench isolation (STI) oxide were still critical challenges in the integration of FinFETs.

We have been able to predict the stress-dependent variability in devices using density-gradient based trigate FinFET device simulation. Due to the presence of RDD and MGG, there is a strong effect on the variability in the on- and off-currents and threshold voltage. Variability induced by the fluctuations in the position and number of the discrete dopants between identically manufactured devices is reported. The impact of location-dependent discrete dopants in different positions such as the source/drain side has been studied to predict the electrical performance. It is shown that RDDs in the source–drain critically affect the performance through the variations in the short channel effects. The variability in on- and off-currents and threshold voltage has been quantified by calculating the mean and standard deviation of the parameters.

A methodology is proposed to simulate the effects of the local WFV due to MGG on device matching. Using this methodology, the threshold voltage distribution will be bound within the extreme values set by the work function. Simulation results show that the process-induced variability sources become important for FinFET devices as they affect more the gate control and the channel charge. Stress profiling analysis is adopted to simulate process steps of SiGe epitaxial S/D p-FinFETs. This study will undoubtedly provide a guideline for trigate FinFET design using stress mapping/tuning which is of great importance for technology scaling.

REFERENCES

[1] E. Karl et al., "*17.1 A 0.6V 1.5GHz 84Mb SRAM design in 14 nm FinFET CMOS technology,*" in *2015 IEEE International Solid-State Circuits Conference - (ISSCC) Digest of Technical Papers*, 2015, pp. 1–3, doi: 10.1109/ISSCC.2015.7063050.

[2] C. Meinhardt, A. L. Zimpeck, and R. A. L. Reis, "Predictive evaluation of electrical characteristics of sub-22 nm FinFET technologies under device geometry variations," *Microelectron. Reliab.*, vol. 54, no. 9, pp. 2319–2324, 2014, doi: 10.1016/j.microrel.2014.07.023.

[3] IEEE, "International roadmap for devices and systems," 2018.

[4] P. H. Vardhan et al., "Analytical modeling of metal gate granularity based threshold voltage variability in NWFET," *Solid. State. Electron.*, vol. 147, pp. 26–34, 2018, doi: 10.1016/j.sse.2018.05.007.

[5] A. Asenov, A. R. Brown, J. H. Davies, and S. Saini, "Hierarchical approach to 'atomistic' 3-D MOSFET simulation," *IEEE Trans. Comput. Des. Integer. Circuits Syst.*, vol. 18, no. 11, pp. 1558–1565, 1999, doi: 10.1109/43.806802.

[6] G. Indalecio et al., "Study of metal-gate work-function variation using Voronoi cells: Comparison of rayleigh and gamma distributions," *IEEE Trans. Electron Devices*, vol. 63, no. 6, pp. 2625–2628, 2016, doi: 10.1109/TED.2016.2556749.

[7] O. Faynot et al., "*Planar fully depleted SOI technology: A powerful architecture for the 20 nm node and beyond,*" in *2010 International Electron Devices Meeting*, 2010, pp. 3.2.1–3.2.4, doi: 10.1109/IEDM.2010.5703287.

[8] K. Xu, R. Patel, P. Raghavan, and E. G. Friedman, "Exploratory design of on-chip power delivery for 14, 10, and 7 nm and beyond FinFET ICs," *Integration*, vol. 61, pp. 11–19, 2018, doi: 10.1016/j.vlsi.2017.10.007.

[9] ITRS Executive Summary - 2015 Edition., International technology roadmap for semi-conductors, 2015.

[10] C. Su, P. Sung, K. Kao, Y. Lee, W. Wu, and W. Yeh, *"Process and structure consider-ations for the post FinFET era,"* in *2020 IEEE Silicon Nanoelectronics Workshop (SNW)*, 2020, pp. 13–14, doi: 10.1109/SNW50361.2020.9131422.

[11] S. Dey, J. Jena, T. P. Dash, E. Mohapatra, S. Das, and C. K. Maiti, *"Performance evalu-ation of gate-all-around Si nanowire transistors with SiGe strain engineering,"* in *2019 IEEE Conference on Modeling of Systems Circuits and Devices (MOS-AK India)*, 2019, pp. 29–33, doi: 10.1109/MOS-AK.2019.8902440.

[12] T. P. Dash, S. Dey, E. Mohapatra, S. Das, J. Jena, and C. K. Maiti, *"Vertically-stacked silicon nanosheet field effect transistors at 3 nm technology nodes,"* in *2019 Devices for Integrated Circuit (DevIC)*, 2019, pp. 99–103, doi: 10.1109/DEVIC.2019.8783300.

[13] S. Reboh et al., "Strain, stress, and mechanical relaxation in fin-patterned Si/SiGe mul-tilayers for sub-7 nm nanosheet gate-all-around device technology," *Appl. Phys. Lett.*, vol. 112, no. 5, p. 51901, 2018, doi: 10.1063/1.5010997.

[14] T. P. Dash, "Design and simulation of strain-engineered trigate FinFETs toward ultimate scaling," PhD Thesis, SOA University, 2019.

[15] T. P. Dash, S. Dey, S. Das, E. Mohapatra, J. Jena, and C. K. Maiti, "Strain-engineering in nanowire field-effect transistors at 3 nm technology node," *Phys. E Low-dimensional Syst. Nanostructures*, vol. 118, p. 113964, 2020, doi: 10.1016/j.physe.2020.113964.

[16] T. P. Dash, J. Jena, E. Mohapatra, S. Dey, S. Das, and C. K. Maiti, "Stress-induced vari-ability studies in tri-gate FinFETs with source/drain stressor at 7 nm technology nodes," *J. Electron. Mater.*, vol. 48, no. 8, pp. 5348–5362, 2019, doi: 10.1007/s11664-019-07348-7.

[17] M. S. Bhoir et al., *"Process-induced Vt variability in nanoscale FinFETs: Does Vt extraction methods have any impact?,"* in *2020 4th IEEE Electron Devices Technology Manufacturing Conference (EDTM)*, 2020, pp. 1–4, doi: 10.1109/EDTM47692.2020.9117815.

[18] A. R. Brown, N. M. Idris, J. R. Watling, and A. Asenov, "Impact of metal gate granular-ity on threshold voltage variability: A full-scale three-dimensional statistical simulation study," *IEEE Electron Device Lett.*, vol. 31, no. 11, pp. 1199–1201, 2010, doi: 10.1109/LED.2010.2069080.

[19] C. C. Hobbs et al., "Fermi-level pinning at the polysilicon/metal oxide interface-Part I," *IEEE Trans. Electron Devices*, vol. 51, no. 6, pp. 971–977, 2004, doi: 10.1109/TED.2004.829513.

[20] X. Wang, G. Roy, O. Saxod, A. Bajolet, A. Juge, and A. Asenov, "Simulation study of dominant statistical variability sources in 32-nm high-κ/metal gate CMOS," *IEEE Electron Device Lett.*, vol. 33, no. 5, pp. 643–645, 2012, doi: 10.1109/LED.2012.2188268.

[21] H. Dadgour et al., *"Statistical modeling of metal gate work-function variability in emerging device technologies and implications for circuit design,"* *Proc. IEEE/ACM Int. Conf. Comput. Des*, pp. 270–277, 2008, doi: 10.1109/ICCAD.2008.4681585.

[22] H. F. Dadgour, K. Endo, V. K. De, and K. Banerjee, "Grain-orientation induced work function variation in nanoscale metal-gate transistors - Part I: Modeling, analysis, and experimental validation," *IEEE Trans. Electron Devices*, vol. 57, no. 10, pp. 2504–2514, 2010, doi: 10.1109/TED.2010.2063191.

[23] T. P. Dash, S. Dey, J. Jena, S. Das, E. Mohapatra, and C. K. Maiti, *"Metal grain granu-larity induced variability in gate-all-around si-nanowire transistors at 1 nm technology node,"* in *2019 Devices for Integrated Circuit (DevIC)*, 2019, pp. 286–290, doi: 10.1109/DEVIC.2019.8783717.

[24] S. M. Nawaz, S. Dutta, and A. Mallik, "A comparison of random discrete dopant induced variability between Ge and Si junctionless p-FinFETs," *Appl. Phys. Lett.*, vol. 107, no. 3, 2015, doi: 10.1063/1.4927279.

[25] K.-C. Lee, M.-L. Fan, and P. Su, "Investigation and comparison of analog figures-of-merit for TFET and FinFET considering work-function variation," *Microelectron. Reliab.*, vol. 55, no. 2, pp. 332–336, 2015, doi: 10.1016/j.microrel.2014.11.012.

[26] K. Nayak, S. Agarwal, M. Bajaj, K. V. R. M. Murali, and V. R. Rao, "Random dopant fluctuation induced variability in undoped channel Si gate all around nanowire n-MOS-FET," *IEEE Trans. Electron Devices*, vol. 62, no. 2, pp. 685–688, 2015, doi: 10.1109/TED.2014.2383352.

[27] H. Nam, Y. Lee, J. Park, and C. Shin, "Study of work-function variation in high-k/metal-gate gate-all-around nanowire MOSFET," *IEEE Trans. Electron Devices*, vol. 63, no. 8, pp. 3338–3341, 2016, doi: 10.1109/TED.2016.2574328.

[28] G. A. Armstrong and C. K. Maiti, *TCAD for Si, SiGe, and GaAs Integrated Circuits*. The Institution of Engineering and Technology (IET), UK, 2008.

[29] Q. Han, M. Liu, B. K. Esfeh, J. H. Bae, J. P. Raskin, and Q. T. Zhao, "Impact of gate to source/drain alignment on the static and RF performance of junctionless Si nanowire n-MOSFETs," *Solid. State. Electron.*, vol. 169, 2020, doi: 10.1016/j.sse.2020.107817.

[30] S. Dey, J. Jena, E. Mohapatra, T. P. Dash, S. Das, and C. K. Maiti, "Design and simulation of vertically-stacked nanowire transistors at 3 nm technology nodes," *Phys. Scr.*, vol. 95, no. 1, p. 14001, 2020, doi: 10.1088/1402-4896/ab4621.

[31] Y. Wang et al., "*Variability-aware TCAD based design-technology co-optimization platform for 7 nm node nanowire and beyond*," in *2016 IEEE Symposium on VLSI Technology*, 2016, pp. 1–2, doi: 10.1109/VLSIT.2016.7573423.

[32] C. S. Smith, "Piezoresistance effect in germanium and silicon," *Phys. Rev.*, vol. 94, no. 1, pp. 42–49, 1954, doi: 10.1103/PhysRev.94.42.

[33] D. Colman, R. T. Bate, J. P. Mize, D. Colman R. T. Bate, and J. P. Mize, "Mobility anisotropy and piezoresistance in silicon p-type inversion layers," *J. Appl. Phys.*, vol. 39, no. 4, pp. 1923–1931, 1968, doi: 10.1063/1.1656464.

[34] GTS Framework, VSP user manual, 2020.

[35] MINIMOS-NT user manual, 2020.

[36] M. Karner, Z. Stanojevic, C. Kernstock, H. W. Cheng-Karner, and O. Baumgartner, "*Hierarchical TCAD device simulation of FinFETs*," in *2015 International Conference on Simulation of Semiconductor Processes and Devices (SISPAD)*, 2015, pp. 258–261, doi: 10.1109/SISPAD.2015.7292308.

[37] Z. Stanojevic, O. Baumgartner, M. Karner, L. Filipovic, C. Kernstock, and H. Kosina, "*Full-band modeling of mobility in p-type FinFETs*," in *2014 Silicon Nanoelectronics Workshop (SNW)*, 2014, pp. 1–2, doi: 10.1109/SNW.2014.7348592.

[38] Z. Stanojevic et al., "*Physical modeling - A new paradigm in device simulation*," in *2015 IEEE International Electron Devices Meeting (IEDM)*, 2015, pp. 5.1.1–5.1.4, doi: 10.1109/IEDM.2015.7409631.

[39] G. Baccarani, E. Gnani, A. Gnudi, S. Reggiani, and M. Rudan, "Theoretical foundations of the quantum drift-diffusion and density-gradient models," *Solid. State. Electron.*, vol. 52, no. 4, pp. 526–532, 2008, doi: 10.1016/j.sse.2007.10.051.

[40] M. Lundstrom, "*Drift-diffusion and computational electronics-still going strong after 40 years!*," in *Simul. Semicond. Process. Devices Conf.*, 2015, doi: 10.1109/SISPAD.2015.7292243.

[41] M. G. Ancona, "Density-gradient theory: A macroscopic approach to quantum confinement and tunneling in semiconductor devices," *J. Comput. Electron.*, vol. 10, no. 1–2, pp. 65–97, 2011, doi: 10.1007/s10825-011-0356-9.

[42] Z. Stanojevic et al., "*Scaling FDSOI technology down to 7 nm? A physical modeling study based on 3D phase-space subband BBoltzmann transport*," in *2018 Joint International EUROSOI Workshop and International Conference on Ultimate Integration on Silicon (EUROSOI-ULIS)*, 2018, pp. 1–4, doi: 10.1109/ULIS.2018.8354741.

[43] S. M. Amoroso, V. P. Georgiev, E. Towie, C. Riddet, and A. Asenov, "*Metamorphosis of a nanowire: A 3-D coupled mode space NEGF study*," in *2014 International Workshop on Computational Electronics (IWCE)*, 2014, pp. 1–4, doi: 10.1109/IWCE.2014.6865854.

[44] G. Wang et al., "Study of SiGe selective epitaxial process integration with high-k and metal gate for 16/14 nm nodes FinFET technology," *Microelectron. Eng.*, vol. 163, pp. 49–54, 2016, doi: 10.1016/j.mee.2016.06.002.

[45] H. Dadgour et al., "Modeling and analysis of grain-orientation effects in emerging metal-gate devices and implications for SRAM reliability," *IEEE Int. Electron Devices Meet.*, no. 1–4, 2008, doi: 10.1109/IEDM.2008.4796792.

[46] H. Nam, C. Shin, and J. Park, "Impact of the metal-gate material properties in FinFET (Versus FD-SOI MOSFET) on high-κ/metal-gate work-function variation," *IEEE Trans. Electron Devices*, vol. 65, no. 11, pp. 4780–4785, 2018, doi: 10.1109/TED.2018.2872586.

[47] S. M. Nawaz, S. Dutta, A. Chattopadhyay, and A. Mallik, "Comparison of random dopant and gate-metal workfunction variability between junctionless and conventional FinFETs," *IEEE Electron Device Lett.*, vol. 35, no. 6, pp. 663–665, 2014, doi: 10.1109/LED.2014.2313916.

[48] M. Choi, V. Moroz, L. Smith, and O. Penzin, "*14 nm FinFET stress engineering with epitaxial SiGe source/drain*," in *2012 International Silicon-Germanium Technology and Device Meeting (ISTDM)*, 2012, pp. 1–2, doi: 10.1109/ISTDM.2012.6222469.

[49] H. V. Swygenhoven et al., "Grain-boundary structures in polycrystalline metals at the nanoscale," *Phys. Rev. B*, vol. 62, no. 2, pp. 831–838, 2000, doi: 10.1103/PhysRevB.62.831.

[50] A. Leonardi et al., "Realistic nano-polycrystalline microstructures: Beyond the classical Voronoi tessellation," *Phil. Mag.*, vol. 92, no. 8, pp. 986–1005, 2012, doi: 10.1080/14786435.2011.637984.

[51] N. Sano, K. Matsuzawa, M. Mukai, and N. Nakayama, "On discrete random dopant modeling in drift-diffusion simulations: Physical meaning of 'atomistic' dopants," *Microelectron. Reliab.*, vol. 42, no. 2, pp. 189–199, 2002, doi: 10.1016/S0026-2714(01)00138-X.

[52] N. Sano et al., "Role of long-range and short-range Coulomb potentials in threshold characteristics under discrete dopants in sub-0.1μm Si-MOSFETs," *IEEE Int. Electron devices Meet.*, pp. 275–278, 2000, doi: 10.1109/IEDM.2000.904310.

[53] S. K. Saha, "Modeling statistical dopant fluctuations effect on threshold voltage of scaled JFET devices," *IEEE Access*, vol. 4, pp. 507–513, 2016, doi: 10.1109/ACCESS.2016.2519039.

7 Technology CAD of III-Nitride Based Devices

The use of alternative channel materials such as III-nitrides featuring high mobility is being currently discussed for inclusion in silicon technology. This is basically under the More-than-Moore applications [1, 2]. Especially, III-V materials or pure Ge have been considered [3] but there are still challenges to be addressed such as the integration on Si substrate or the leakage induced by band-to-band tunneling for instance. Strain engineering remains one of the most powerful boosters to increase the performance of CMOS technology [4]. It is indeed today being discussed for achieving sub-10 nm node requirements. More details about the strain integration in CMOS technology have been presented in Chapter 3.

Gallium nitride (GaN) nanoelectronics have operated at temperatures as high as 1,000°C making it a viable platform for robust space-grade electronics and nanosatellites. Also, there has been a tremendous amount of research and industrial investment in GaN as it is positioned to replace silicon in the billion-dollar power electronics industry, as well as the post-Moore microelectronic applications. Furthermore, the 2014 Nobel Prize in physics was awarded for pioneering research in GaN that led to the realization of the energy-efficient blue light-emitting diodes (LEDs). Even with these major technological breakthroughs, we have just begun the "GaN revolution." New communities are adopting this nanoelectronic platform for a multitude of emerging device applications including the following: sensing, energy harvesting, actuation, and communication. In this chapter, we will review and discuss the benefits of GaN's 2D electron gas over silicon's p-n junction. Also, we will discuss recent results that advance this nanoelectronic device platform for Internet-of-Things (IoT) systems.

In power electronics, different semiconductors can be considered as silicon (Si), gallium arsenide (GaAs), indium phosphide (InP), carbide silicon (SiC), element III-nitrides (GaN, AlN, InN), and diamond. Although the silicon sector dominates, thanks to its low cost and its perfectly mastered technology, this sector has reached its limits. The margin for improvement of silicon components is reduced and the use of other materials allows them to be overcome. Table 7.1 lists the main physical and electrical properties of these different semiconductor materials. The high bandgap allows for a high breakdown field, which is interesting for power applications. From Table 7.1, GaN shows superiority over GaAs and low bandgap semiconductors. Also, AlN and diamonds are very good candidates and offer a lot of potentials, only the availability in reasonable size with well-controlled electrical properties appears to be difficult, which limits its development. Also, GaN has a high saturation rate desirable for high-frequency operation. To achieve high current densities, the semiconductor material must also exhibit high electronic mobility. Although the mobility value is relatively low in GaN compared to that found in GaAs, it is sufficient for high-power

TABLE 7.1

Intrinsic Properties of the Various Semiconductor Materials at 300 K

Properties	Si	GaAs	InP	4H-SiC	GaN	AlN	InN	C-Diamond
Band energy forbidden Eg (eV)	1.1	1.42	1.35	3.25	3.4	6.2	0.7	5.5
Breakdown field (MV/cm)	0.3	0.4	0.5	3	3.3	8.4	1.2	5.6
Velocity saturation ($\times10^7$ cm/s)	1	1.3	1	2	2.5	2.1	1.8	2.7
Mobility of electrons (cm²/V.s)	1,500	8,500	5,400	800	900/2,000*	300	3.6	1,800
Thermal conductivity (W/cm.K)	1.5	0.5	0.7	4.5	1.7	2	1.8	20
Relative permittivity	11.8	11.5	12.5	10	9.5	9.14	15.3	5.7

* Heterostructure AlGaN/GaN

operation. It is noted that the density of electrons in the channel in the case of a GaN-based heterojunction is significantly higher; it is of the order of 10^{13} cm^{-2}. Furthermore, thermal conductivity is an important parameter because it determines the heat that can be dissipated by the component. The increase in heat limits the operation at a very high temperature, so it is necessary to have devices with very good thermal conductivity. The value for GaN is three times larger than that for GaAs and almost similar to that of Silicon, but it is notably lower than that of SiC, AlN, and diamond.

For a better comparison of performance in power and frequency potentially accessible with these different semiconductor materials, figures of merit have been proposed. Most used in the field of power electronics and high frequency is the Johnson figure of merit (JFM) and the Baliga figure of merit (BFM). Johnson's figure of merit takes into account the breakdown field and the speed of saturation of electrons, while the BFM involves the mobility of electrons, the relative permittivity as well as the breakdown field. Table 7.2 groups together the values of the main figures of merit normalized concerning silicon. The GaN has a fairly high potential with a good compromise between power and frequency high. Also, materials such as diamond and

TABLE 7.2

Figure of Merit of the Various Semiconductor Materials (Normalized to Silicon)

	Si	GaAs	InP	4H-SiC	GaN	AlN	InN	C-Diamond
Figure of merit of Johnson ($E_{br}.v_{sat}/2\pi^2$	1	3	2.8	400	760	3,457	52	2,540
Figure of merit of Baliga ($\mu_e.\varepsilon_r.E_{br}^3$)	1	13	18	452	643	3,400	0.2	3,770

AlN have a higher potential than GaN, but research is not yet advanced enough for these components to be exploited. The realization of HEMT type structures based on AlN seems difficult because of a strong lattice mismatch with the other III-V compounds. On the other hand, it has been shown that the use of AlN as a very thin barrier layer in heterostructures based on GaN improves the mobility of electrons in the channel [5, 6]. Diamond is often described as the ideal semiconductor for power electronics applications. Of more, used as a substrate it presents an improvement of the performances by reducing the thermal resistance of components [7]. However, the growth of diamond on a substrate host remains limited because of the strong strain generated.

7.1 HISTORY OF NITRIDE TECHNOLOGY

Although the synthesis of GaN dates from the 1920s and 1930s [8], the real interest in nitrides III elements began in the late 1960s. Maruska et al. [9] have shown the possibility of growing GaN single crystals on a sapphire substrate. However, due to the difficulties encountered (high dislocation density, high n-type residual doping, etc., and also the low p-doping efficiency), research on nitrides has almost stopped. In 1986, Amano et al. managed to grow a GaN layer with morphological properties with greatly improved optical and electrical [10]. The quality of GaN has been improved by growth by the Metal-Organic Chemical Vapor Deposition (MOCVD) technique on a sapphire substrate using an AlN buffer layer. A few years later, the same group of researchers obtained efficient p-type doping with magnesium (Mg), which led to the first GaN-based p-n junction LED [11]. The 1990s saw some progress in the growth of good quality GaN allowing the rebirth of this technology. In 1991, the first observation of a 2D electron gas at the Aluminum Gallium Nitride/ Gallium Nitride (AlGaN/GaN) heterojunction was reported by Khan et al. [12]. Two years later, this progress has led to the first heterostructure-based high electron mobility transistor (HEMT) using AlGaN/GaN produced by MOCVD on a sapphire substrate [13]. In the domain of optoelectronics, Nakamura et al. [14] reported the first blue LED, and 2 years later, with the first GaN-based blue laser [15]. In 2014, Akasaki, Amano, and Nakamura received the Nobel Prize for their work leading to efficient blue LEDs. Today's most successful applications of GaN are LEDs for optoelectronic applications and transistors with high electron mobility for electronic applications. GaN is indeed a very good candidate for power electronics since it can withstand high power densities and large voltages of the breakdown in comparison with other III-V compounds while maintaining frequencies of high operation.

7.2 MATERIAL PROPERTIES III-N

III-nitrides are semiconductor materials formed by covalent bonding of an element from column III such as Gallium (Ga), Aluminum (Al), Indium (In), and the element Nitrogen (N) from column V of the periodic table. Binary nitrides occur mainly in two crystalline forms: the Zinc structure blende (cubic phase) and the wurtzite structure (hexagonal phase). The crystal structure is related to the growing conditions as

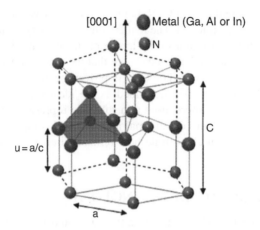

FIGURE 7.1 Wurtzite crystal structure of III-N. Atomic structure of GaN as wurtzite crystal. Ga atoms are represented in blue color and are relatively larger than N atoms represented in gray. The bonds are presented in two colors. Lattice parameters along with bond length are presented in blue arrows.

well as to the surface symmetry of the substrate. The zinc blende structure is obtained by growth under non-equilibrium conditions thermodynamics (low temperature) from the (001) planes of cubic substrates. The wurtzite structure is the most commonly used because it is the most stable thermodynamically. The ionic nature of the III-N bond provides great stability to the structure, due to the high electronegativity of nitrogen.

Figure 7.1 represents the lattice elementary of a wurtzite structure defined by the three lattice parameters: a (length of the base hexagon), c (height of the elementary cell) which corresponds to the distance between two atoms of the same nature, and u (bond length between atoms III and N). The wurtzite structure consists of two hexagonal subnetworks, one occupied by element III and the other by nitrogen (N), interpenetrated and shifted along axis [0001] by 5/8 of the cell unit. In an ideal wurtzite crystal, the relation between these three parameters should be u = 3/8 and c/a = 1.633 which results in regular tetrahedra (each atom is linked to four atoms neighbors, three of which happen to be in the same crystallographic plane perpendicular to the axis). However, the III-N wurtzite structure deviates slightly from the ideal wurtzite structure. So, the element III atoms are in a somewhat distorted tetrahedral environment. Therefore, the sum of the dipole moments is non-zero considering that the atom of element III has a positive charge, and the more electronegative nitrogen atom presents a negative charge.

The wurtzite structure does not have an inversion center (not centro-symmetrical), which means that the two crystalline directions [0001] and [000-1] are not equivalent. This structure is said to be polar; we thus find a structure of metal polarity associated with the bond III-N oriented along the crystallographic axis [0001] and a structure of nitrogen polarity corresponding to the III-N link oriented along the [000-1] axis (see Figure 7.2).

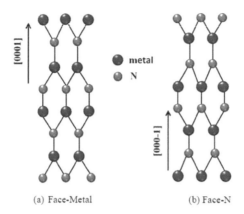

(a) Face-Metal   (b) Face-N

FIGURE 7.2   Wurtzite crystal structures of GaN: a) metal-face [0001], b) N-face [000-1].

These different polarities have properties that affect both the technology and material properties, due to critical growing conditions. Indeed, the polarity influences the crystal surface properties, morphology, and chemical stability. III-N materials at nitrogen polarity suffer from a rougher surface morphology, which favors the incorporation of impurities during growth and penalizes resistivity and electrical insulation of the buffer layer of material. It, therefore, follows that the metal polarity is the most frequently used for the majority of applications. Recently, element III-nitrides with non-polar and semi-polar surface orientations have attracted much attention due to the possibility of reducing the concentration of defects and avoiding or reducing the appearance of internal electrical polarization. The considerations of crystalline orientation have great importance because they determine the orientation of the spontaneous and piezoelectric polarization in the surface and material interface of a heterostructure [16, 17]. The values of the lattice parameter a, c, and u for materials III-N are shown in Table 7.3.

Concerning ternary alloys (AlGaN, InAlN, and InGaN), the lattice parameters obey a linear Vegard's law according to the composition x:

$$Y\left(AxB1-xN\right) = xY\left(AN\right) + \left(1-x\right)Y\left(BN\right) \tag{7.1}$$

TABLE 7.3
a and c Lattice Parameters and Spontaneous Polarization
Coefficients of Elemental III-Nitrides

Wurtzite	$a(\text{Å})$	$c(\text{Å})$	$u(\text{Å})$	$P_{sp}(C/m^2)$
GaN	3.189	5.185	0.376	−0.034
AlN	3.112	4.982	0.380	−0.090
InN	3.540	5.705	0.377	−0.042

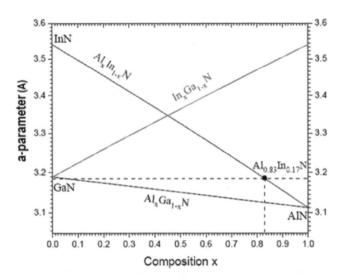

FIGURE 7.3A Lattice parameter a (Å), dependence on the alloy composition. The InAlN alloy with a composition of Al = 0.83 has the same a-lattice parameter as GaN.

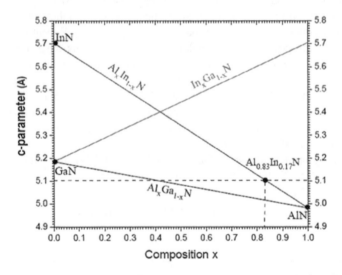

FIGURE 7.3B Lattice parameter c (Å) dependence on the alloy composition. The InAlN alloy with a composition of Al = 0.83 has the same a-lattice parameter as GaN.

where A and B represent Al, Ga, or In and Y represents the lattice parameters. Figures 7.3a and 7.3b represent the relationship between the composition of an alloy and its lattice parameters a and c. As can be seen in Figure 7.3a, the InAlN alloy with a 17% indium composition has the advantage of having the same lattice parameter as GaN. In other words, it is lattice-matched with GaN in the growth plane [0001]. This has the effect of canceling the stresses and consequently produces zero piezoelectric polarization.

7.3 OPTICAL PROPERTIES

The bandgap represents the energy required for an electron to pass from the band valence to the conduction band. The width of the forbidden band or gap measures the energy gap between the maximum of the valence band and the minimum of the conduction band. The direct bandgap energy of element III-nitrides ranges from 0.7 eV for InN to 3.4 eV for GaN and reaches 6.2 eV for AlN. The possibility of easily forming alloys makes it possible to modulate the width of the strip prohibited depending on the composition of the alloy. Bandgap energies of alloys are commonly described by the following equation:

$$E_g^{A_x B_{1-x} N} = x E_g^{AN} + (1-x) E_g^{BN} - b^{ABN} x (1-x) \qquad (7.2)$$

where b is the term of non-linearity (bowing), it represents the deviation compared to the linear law. It is 1.0 eV for AlGaN [16], 1.32 eV for InGaN [18] and 5.36 eV for InAlN. For the alloy InAlN, the value of 5.36 eV proposed by Sakalauskas et al. [19], is in good agreement with the experimental values of the forbidden bands. However, a more precise study showed a dependence on the non-linearity term on the composition of indium [20].

$$b^{InAlN}(x) = \frac{A}{1 + Cx^2} \qquad (7.3)$$

where $A = 6.43 \pm 0.12$ eV and $C = 1.21 \pm 0.14$ eV.

Another expression was introduced by Ilopoulos et al. [21]:

$$b^{InAlN}(x) = \frac{A_1}{1 + A_2 x^2} \qquad (7.4)$$

where $A_1 = 15.3 \pm 1.6$ eV and $A_2 = 4.81 \pm 0.95$ eV.

Figure 7.4 shows one of the advantages of III-N materials is the possibility of fully covering the spectral range beyond the visible range, with a broad spectrum wavelength ranging from infrared (1.85 μm) to deep ultraviolet (200 nm) by variation of the composition of alloys. However, the feasibility of certain compositions remains limited for reasons notably related to the management of the strain generated by the heteroepitaxy of III-nitrides, or the large differences in the binding energy of compounds in some alloys like InAlN. Another parameter to take into account is the variation in the energy of the gap as a function of the temperature. It is very well known that due to thermal agitation in the network crystalline, the energy of the

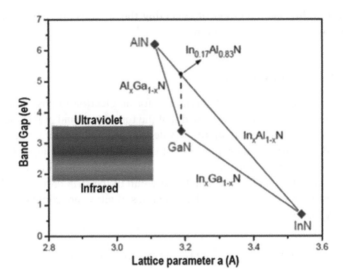

FIGURE 7.4   Bandgap energy as a function of the lattice parameter of III-N materials.

bandgap decreases slightly with increasing temperature. This behavior can be expressed as:

$$E_g(T) = E_g(0) - \frac{\alpha T^2}{T + \beta} \tag{7.5}$$

where $E_g(0)$ corresponds to the energy of the forbidden band at $T = 0$ K, α (meV/K) represents the coefficient linear regression of the gap for high temperatures and β (K) indicates the temperature where there is a change in slope.

7.3.1   Crystal Deformation: Biaxial and Uniaxial Case

The pseudomorphic growth of a film with a different lattice parameter than the substrate imposes a deformation and the appearance of stress in the plane of growth takes place. The stress and strain of the film are linked by the laws of linear elasticity (Hooke's law). The crystal deformation (strain) of the epilayer can be related to the stress it suffered from the fact that the in-plane lattice parameter of the epilayer changed to that of the substrate lattice parameter (a-parameter). Moreover, the out-of-plane lattice parameter (c-parameter) of the epilayer changes depending on the in-plane strain induced during the growth. The stress–strain relation and also the relationship between the out-of-plane lattice parameter (c-parameter) to that of the in-plane lattice parameter (a-parameter) can be found from Hooke's law which allows us to relate the crystal stress to that of the crystal strain as follows:

$$\vec{\sigma} = C.\vec{\varepsilon} \tag{7.6}$$

where $\vec{\sigma}$ and $\vec{\varepsilon}$ are stress and strain tensors and C is the elastic coefficient tensor. Using Voigt notation for hexagonal crystal system, we can express them as follows:

$$\vec{\sigma} = \begin{bmatrix} \sigma_{xx} \\ \sigma_{yy} \\ \sigma_{zz} \\ \sigma_{yz} \\ \sigma_{zx} \\ \sigma_{xy} \end{bmatrix} \vec{\varepsilon} = \begin{bmatrix} \varepsilon_{xx} \\ \varepsilon_{yy} \\ \varepsilon_{zz} \\ 2\varepsilon_{yz} \\ 2\varepsilon_{zx} \\ 2\varepsilon_{xy} \end{bmatrix} C = \begin{bmatrix} C_{11} & C_{12} & C_{13} & 0 & 0 & 0 \\ C_{12} & C_{11} & C_{13} & 0 & 0 & 0 \\ C_{13} & C_{13} & C_{33} & 0 & 0 & 0 \\ 0 & 0 & 0 & C_{44} & 0 & 0 \\ 0 & 0 & 0 & 0 & C_{44} & 0 \\ 0 & 0 & 0 & 0 & 0 & \dfrac{C_{11}-C_{12}}{2} \end{bmatrix} \quad (7.7)$$

If we consider that the growth direction of an epilayer is along Z-axis, then the stress in the grown epilayer will take place along the two perpendicular directions of the in-plane axes (along X and Y axes). In such a case, the heterostructure is known to be in a biaxial stress state. In general, the strain tensor components along different directions are expressed in terms of the lattice constants as:

$$\varepsilon_{xx} = \varepsilon_{yy} = \frac{a-a_0}{a_0}; \varepsilon_{zz} = \frac{c-c_0}{c_0} \quad (7.8)$$

where a_0 and c_0 are relaxed epilayer's unit cell parameters and a and c are the strained epilayer's unit cell parameters. The absence of stress along growth direction ($\sigma_{zz} = 0$) and considering that the strain is equal in two lateral directions, impose the correlation between strain and stress from Equations (7.6, 7.7) as:

$$\sigma = \left(C_{11} + C_{12} - \frac{2c_{13}^2}{c_{33}} \right) \varepsilon \quad (7.9)$$

where $\sigma = \sigma_{xx} = \sigma_{yy}$, and $\forall_i \neq j, \sigma_{ij} = 0$
and

$$\varepsilon_{zz} = -\frac{2c_{13}}{c_{33}} \varepsilon \quad (7.10)$$

where $\varepsilon = \varepsilon_{xx} = \varepsilon_{yy}$

Therefore, for the biaxial stress condition, Equation (7.9) gives us the stress–strain relation of the epilayer, and Equation (7.10) correlates the extension or contraction along the growth direction ε_{zz} (Z-axis) of the epilayer to that of any lateral deformation ε.

In the case of an epilayer with a very thin lateral thickness, the strain along the growth axis, ε_{zz}, may not follow Equation (7.10), as the two lateral strains ε_{xx} and ε_{yy} may not be equal or even one may become zero due to surface relaxation effect along the thin thickness direction. The situation when the strain along the thin lateral thickness is zero is known as the uniaxial stress condition. In uniaxial stress condition (i.e., $\sigma_{xx} \neq \sigma_{yy} = \sigma_{zz} = 0$), the correlation between ε_{xx} and ε_{zz} can be expressed from Equations (7.6, 7.7) as:

$$\sigma_{xx} \neq \sigma_{yy} = \sigma_{zz} = 0 \qquad (7.11)$$

$$\varepsilon_{zz} = -\frac{c_{13}(c_{11} - c_{12})}{c_{11}c_{33} - c_{13}^2}\varepsilon_{xx} \qquad (7.12)$$

For an InGaN/GaN pseudomorphic heteroepitaxial growth process, from Equations (7.10, 7.12), we can express the deformed lattice parameter of any InGaN epilayer along the growth direction as:

$$c = c_0\left[1 - \frac{2c_{13}}{c_{33}}\left(\frac{a - a_0}{a_0}\right)\right](\text{Biaxial stress condition}) \qquad (7.13)$$

$$c = c_0\left[1 - \left\{\frac{c_{13}(c_{11} - c_{12})}{c_{11}c_{33} - c_{13}^2}\right\}\left(\frac{a - a_0}{a_0}\right)\right](\text{Uniaxial stress condition}) \qquad (7.14)$$

The change in lattice parameter between the InGaN epilayer and that of the GaN substrate layer along the growth direction (c-parameter) can be expressed as:

$$\Delta c = c - c_{GaN} \qquad (7.15)$$

Thus the percentage of the InGaN lattice deformation (e_{zz}) along the growth direction to that of GaN substrate lattice parameter (c_{GaN}) can be expressed as:

$$e_{zz}(\%) = 100 \times \frac{c - c_{GaN}}{c_{GaN}} \qquad (7.16)$$

Using Equations (7.13, 7.14), we can determine the percentage of InGaN lattice deformation for both biaxial and uniaxial stress condition from Equation (7.16) as:

$$e_{zz}(\%) = 100 \times \left[\frac{c_0}{c_{GaN}}\left[1 - \frac{2c_{13}}{c_{33}}\left(\frac{a - a_o}{a_0}\right)\right] - 1\right](\text{Biaxial stress condition}) \qquad (7.17)$$

$$e_{zz}(\%) = 100 \times \left[\frac{c_0}{c_{GaN}}\left[1 - \left\{\frac{c_{13}(c_{11} - c_{12})}{c_{11}c_{33} - c_{13}^2}\right\}\left(\frac{a - a_o}{a_0}\right)\right] - 1\right](\text{Uniaxial stress condition})$$

$$(7.18)$$

By definition, in pseudomorphic growth, the in-plane lattice parameter of the strained InGaN layer (a) takes the same lattice parameter value as that of relaxed

TABLE 7.4

Elastic and Piezoelectric Coefficients of Element III-Nitrides [22]

Wurtzite	C_{11} (GPa)	C_{12} (GPa)	C_{13} (GPa)	C_{33} (GPa)	C_{44} (GPa)
GaN	367	135	103	405	95
AlN	396	137	108	373	116
InN	223	115	92	224	48

GaN, that is, $a = a_{0GaN}$. Moreover, the relaxed parameters (a_0 and c_0) of the InGaN layer within these Equations (7.13) to (7.18) depending on the Indium composition (x) and can be determined from Vegard's law and the unstrained lattice parameters of GaN and InN materials ($c_{0GaN} = 5.185$ Å, $a_{0GaN} = 3.189$ Å, $c_{0InN} = 5.718$ Å, $a_{0InN} = 3.544$ Å). Due to the symmetry of the hexagonal network, the elasticity tensor is given by five parameters: C_{11}, C_{12}, C_{13}, C_{33}, and C_{44}, which are reported in Table 7.4.

If we consider only epitaxy in the growth plane, the structure undergoes biaxial stress, and therefore the component of the stress along with the c-axis is absent ($\sigma_3 = 0$) thus that the shear components ($\sigma_4 = \sigma_5 = \sigma_6 = 0$), while the components σ_1 and σ_2 are not zero. The biaxial stress also causes a deformation in the growth plane and along with the c-axis. In most of the calculations performed, the growth is assumed to be pseudomorphic. However, this assumption is not always realistic depending on the structure and the substrate. Due to these deformations, the film must store elastic energy. Beyond a certain thickness known as critical thickness, it will be energetically more favorable to relax the energy elastic by plastic deformation creating dislocations or cracks. This must be taken into account for the optimization of growth to avoid relaxation or the formation of cracks.

7.4 POLARIZATION IN NITRIDE SEMICONDUCTORS

As mentioned earlier, the big difference in electronegativity between the metal and the nitrogen atom causes a transfer of charges in the structure, more important near the N atom which is more electronegative. The wurtzite phase structure is not ideal ($c/a = 1.633$), so the barycenter of positive and negative charges do not coincide. This phenomenon is at the origin of the existence of spontaneous polarization (P_{sp}). The vector (P_{sp}) which results, is oriented along with the c-axis in the direction [000-1]. Indeed, this vector is oriented in the opposite direction to the direction of growth in the case of a metal polarity while that it is in the same direction for a nitrogen polarity (see Figure 7.5). By convention, spontaneous polarization strongly depends on the difference in electronegativity between atoms and the c/a ratio.

The value of the spontaneous polarization of ternary alloys cannot be extrapolated by linear interpolation. Ambacher and Bernardini [23] presented nonlinear laws by adding to the expression of spontaneous polarization a term of bowing." This term represents the contribution of the volume deformation of the binaries and the effects of the internal stress of the alloy, it depends on the lattice parameter u in c-direction.

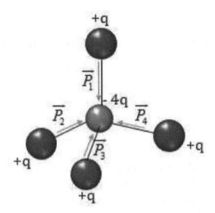

FIGURE 7.5 Spontaneous polarization and charge distribution in an III-N wurtzite structure.

Spontaneous polarization is written therefore from the known values for GaN, AlN, and InN by the following empirical relations:

$$P_{SP}^{Al_xGa_{1-x}N} = -0.090x - 0.034(1-x) + 0.037x(1-x)$$
$$P_{SP}^{In_xAl_{1-x}N} = -0.042x - 0.090(1-x) + 0.070x(1-x) \qquad (7.19)$$
$$P_{SP}^{In_xGa_{1-x}N} = -0.042x - 0.034(1-x) + 0.037x(1-x)$$

It may be noted that the value of the spontaneous polarization of III-N alloy is always negative. The negative sign of the polarization indicates that the vector of polarization points to the surface opposite to the direction [0001]. Thus, the total polarization in strain group III nitride crystals can be expressed as the sum of spontaneous and piezoelectric polarization as follows:

$$P = P_{pz} + P_{sp} \qquad (7.20)$$

In the wurtzite nitride structure, strong polarization (P) exists along with the c-axis [0001]. This polarization is the sum of two types of polarizations known as spontaneous (P_{sp}) and piezoelectric (P_{pz}) polarization. Due to the non-centrosymmetric property and also the difference in electronegativity of the metal atom to that of the nitrogen atom within the wurtzite nitride structure, small electric dipoles exist along c-axis [0001]. These electric dipoles create spontaneous polarization within the structure. Moreover, under stress conditions, the elementary tetrahedrons of the wurtzite structures are deformed which causes the barycenters of the positive and negative charges of the crystal to move away from their usual location, creating an additional electric dipole. Polarization from such induced electric dipole is termed piezoelectric polarization (P_{pz}).

This macroscopic polarization P_{pz} can be defined as a function of the stress as follows:

$$Ppz = D.\vec{\sigma} \qquad (7.21)$$

where D is the piezoelectric tensor. Considering the wurtzite symmetry, the above equation can be written using the Voigt notation as:

$$\begin{bmatrix} p_x^{pz} \\ p_y^{pz} \\ p_z^{pz} \end{bmatrix} = \begin{bmatrix} 0 & 0 & 0 & 0 & d_{15} & 0 \\ 0 & 0 & 0 & d_{15} & 0 & 0 \\ d_{31} & d_{31} & d_{33} & 0 & 0 & 0 \end{bmatrix} \begin{bmatrix} \sigma_{xx} \\ \sigma_{yy} \\ \sigma_{zz} \\ \sigma_{yz} \\ \sigma_{zx} \\ \sigma_{xy} \end{bmatrix} \tag{7.22}$$

Moreover, the piezoelectric polarization can also be expressed as the function of strain tensor as follows:

$$P_{pz} = E \cdot \vec{\varepsilon} \tag{7.23}$$

$$\text{with } E = \begin{bmatrix} 0 & 0 & 0 & 0 & e_{15} & 0 \\ 0 & 0 & 0 & e_{15} & 0 & 0 \\ e_{31} & e_{31} & e_{33} & 0 & 0 & 0 \end{bmatrix} \tag{7.24}$$

The typical value of the piezoelectric coefficients [24] and the spontaneous polarization for III-nitrides are shown in Table 7.5. It may be noted that piezoelectric polarization exhibits a nonlinear behavior concerning strain. This phenomenon has to be taken into account to fit with experimental works. It is worth noticing that those coefficients are significantly larger in group III nitrides than the values reported for conventional III-V semiconductors and also comparable to the values reported for group II-VI wurtzite oxides (ZnO) so that the piezoelectric fields observed in nitrides are very intense.

In the case of heterostructures comprising a thin layer (AlGaN, InAlN, or InGaN) epitaxy on a thick relaxed layer, the ternary alloys undergo extensive deformations, thus, spontaneous and piezoelectric polarizations have the same sign. The effect of the inverse is observed when the stresses are compressive. In the particular case of the InAlN alloy, the composition of indium determines the nature of the strain. For an indium composition greater than 17%, the stress is in compression and therefore

---

TABLE 7.5

Piezoelectric Coefficients and Spontaneous Polarization Values for AlN, GaN, and InN

	e_{31} (C/m²)	e_{33} (C/m²)	d_{31} (pm/V)	d_{33} (pm/V)	d_{15} (pm/V)	P_{sp} (C/m²)
AlN	−0.6	1.46	−2.1	5.4	2.9	−0.081
GaN	−0.49	0.73	−1.4	2.7	1.8	−0.029
InN	−0.57	0.97	−3.5	7.6	5.5	−0.032
ZnO	−0.51	0.89	−5.1	12.4	-	−0.057

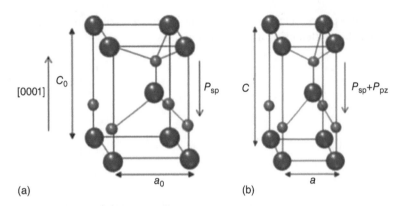

FIGURE 7.6   Representation of the elementary cell and the direction of the polarization in the case (a) without stress and (b) compressive stress.

piezoelectric polarization opposes spontaneous polarization. A composition in 17% indium gives a lattice agreement with GaN, thus a piezoelectric polarization almost zero. It should be noted that the piezoelectric polarization is zero for an unstrained material. The direction of the polarization in the case a) without stress and b) compressive stress is shown in Figure 7.6.

Finally, it may be noted that element III-nitrides have better properties than those of III-V materials (GaAs, InP, etc.), and therefore they are necessary for better piezoelectricity than other materials. Indeed, spontaneous polarization and piezoelectric have significant effects on the electronic properties of heterostructures based on III-N materials. This results in a polarization discontinuity at the interface of a heterojunction that will create an electric field with values up to several MV/cm. To explain the formation of the 2D gas of electrons in nitrides of Element III, GaN-based heterostructures will be used as an example. In this example, a ternary alloy (barrier layer) is deposited during growth pseudomorphic, on a thick GaN layer (channel). Both layers are no intentionally doped. The difference in the lattice parameter between the barrier layer and the GaN channel generates strain which will subsequently modify the polarization by its piezoelectric component. To maintain charge neutrality across the barrier, the net surface charge must be positive and therefore a negative compensating charge is required. Since the layer GaN is of Gallium polarity, the polarization charge at the interface is positive and serves to attract high electron densities. It, therefore, results in the generation of a 2D electron gas at the barrier/channel interface whose electron density can reach 10^{13} cm^{-2} without necessarily resorting to doping the structure.

Khan et al. [25] first demonstrated in 1993 that the presence of electron gas 2D in the AlGaN/GaN heterostructure. At this time, the origin of the 2D electron gas was still under discussion due to the low level of residual doping. However, the model proposed by Ibbetson et al. [26] is the most reasonable and explains relevant to the origin of electrons (see Figure 7.7). This model suggests that the electrons come from donor states deep at the surface of the barrier. Surface states do not ionize only when the barrier thickness exceeds a critical thickness. When the barrier is too fine, the

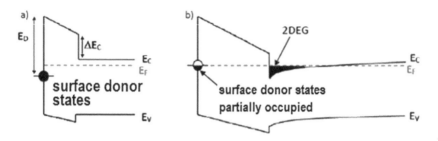

FIGURE 7.7 Schematic illustration of the surface donor model for a barrier thickness (a) less and (b) greater than the critical thickness necessary for the formation of 2D gas.

surface states are below the Fermi level, where the probability of occupancy is maximum. There is therefore no ionization of the surface atoms and the 2D gas is nonexistent. Beyond a certain thickness, the polarization effects will raise the donor states above the Fermi level, electrons can be transferred toward the interface where the containment allows the formation of 2D gas.

The presence of a polarization induced by the interface charges has a strong impact on the band structure of the III-N heterostructure. The principle of 2D electron gas formation then relies on this polarization at the interface. Discontinuities in conduction bands and valence appear at the heterointerface due to the difference in bandgap energy between the barrier and the GaN buffer layer. These discontinuities, therefore, lead to the formation of a potential well on the GaN side where the electronic carriers are confined.

Mobility is a crucial parameter that characterizes the transport of carriers (2D gas) subjected to an electric field and reflects the crystalline quality of the heterostructure. Electrons are driven with a so-called drift speed, which is proportional to the electric field applied. At a low electric field, mobility is independent of the applied field and the drift speed evolves linearly with it. Afterward, the speed shows a peak that corresponds to the maximum speed. When the electric field increases, mobility is no longer constant and varies greatly, the drift speed reaches a saturation level which is due to interactions of carriers with the crystal lattice via optical phonons (mechanism of diffusion predominantly at the strong field).

To explain the formation of the 2D gas of electrons in nitrides of Element III, GaN-based heterostructures consisting of a ternary alloy (barrier layer) is deposited during growth pseudomorphic, on a thick GaN layer (channel). Both layers are no intentionally doped. The difference in the lattice parameter between the barrier layer and the GaN channel generates strain which will subsequently modify the polarization by its piezoelectric component. According to Maxwell's laws, the variation in polarization is manifested by an accumulation of surface charges s_{pol} at the interfaces, given by the gradient of the total polarization $P_{tot.}$

To maintain charge neutrality across the barrier, the net surface charge must be positive and therefore a negative compensating charge is required. Since the layer GaN has Gallium polarity, the polarization charge at the interface is positive and serves to attract high electron densities. It, therefore, results in the generation of a 2D electron gas at the barrier/channel interface whose electron density can reach 10^{13}

cm^{-2} without necessarily resorting to doping the structure. For low carrier densities, mobility is largely limited by dislocations and residual impurities. When N$_s$ increases, the mobility of electrons follows a bell type curve and the mechanism of diffusion by the interface roughness takes precedence over the rest.

The introduction of mobility makes it possible to define sheet resistance. The sheet resistance is a figure of merit for 2D gas, whose value depends on the electron density and the electron mobility μ. The sheet resistance is an important parameter for the correct operation of the transistor; it conditions the value of the access resistors and therefore the current which can pass through the channel, the transconductance, and cutoff frequencies.

7.5   HEMT STRUCTURES

The HEMT structure based on nitride materials using a GaN channel and the barrier made of AlGaN or InAlN to create the heterojunction is the most common. It is constituted a stack of three elements comprising the substrate, a buffer layer comprising the channel, and a barrier layer (see Figure 7.8). For a better understanding of the role of each of the layers of the structure as well as their impact on transistor performance, a description of these different layers is detailed below.

7.5.1   BUFFER LAYER

The quality of a structure is defined both in terms of insulation and the crystalline quality of the buffer layer. This layer must therefore make it possible to obtain a channel, on the one hand, of good structural quality (minimum dislocations, low interface roughness), and also of good electrical quality with a minimum of traps and ionized impurities to avoid leaks in the buffer and the diffusion of electrons, harmful for the transport properties. However, the growth of GaN on an oriented silicon substrate (111) introduces defects that make it difficult to develop heterostructures with the desired criteria. In this case, the use of intermediate layers (complex buffer layers) is a step necessary to manage the strain and filter the defects in the buffer layer generated during growth. These layers consist of a stack of layers of AlN, GaN, and

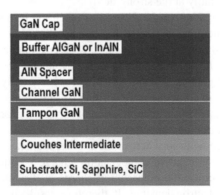

FIGURE 7.8   Schematic description of a HEMT heterostructure.

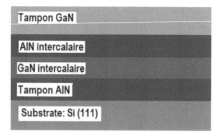

FIGURE 7.9   Structure of buffer layers typically used in GaN HEMTs.

AlN. Figure 7.9 shows the stack of intermediate layers used for structure generation. An AlN nucleation layer of a few nanometers is deposited on the Si substrate. This layer ensures a gradual transition of the lattice parameter between the Si and the GaN layer. Also, it prevents the chemical attack of the substrate surface by gallium and results in the lowest possible dislocation densities. Then a stack of thin layers AlN/GaN also called interlayer is deposited. Finally, a buffer layer of relatively thick, crack-free GaN can be obtained.

The active layers of a HEMT heterostructure consist of a GaN channel, a spacer AlN, a barrier, and a protective layer (cap layer). A GaN cap layer with a thickness varying between 1 nm and a few nanometers, allows protecting the surface of the barrier layer throughout the manufacturing process of the transistor. Finishing the structure with a GaN cap rather than the barrier helps to improved Schottky contact. Finally, its presence makes it possible to enhance the effective potential of the barrier, thanks to the polarization fields present at the interface and therefore to limit reverse bias gate leakage currents [27]. The barriers used are the AlGaN and InAlN barriers. As discussed before, obtaining 2D gas results from the difference in polarization between the GaN and a barrier. However, the formation of 2D gas at the interface with good properties of transport requires that the barrier be granted in the lattice parameter (AlInN with 17% indium) or pseudomorphic on the buffer layer. For this, the thickness of the barrier must be less than the critical relaxation thickness [28]. The insertion of a thin AlN layer called a spacer helps reduce the penetration of electrons in the barrier and to limit the diffusion due to the alloy disorder in favor of better mobility of electrons [29]. This can be explained by the width of the band high forbidden of AlN (6.2 eV) which makes it possible to increase the discontinuity of the conduction (>1.7 eV) between the barrier and the channel.

7.6   SIMULATION CASE STUDIES

7.6.1   STRESS-ENGINEERED AlGaN/GaN HEMT DESIGN

The excellent physical properties of GaN and alloy AlGaN make the devices based on an AlGaN/GaN heterojunction the most interesting candidate for the next generation of power electronics, in particular for space applications. The unique feature of AlGaN/GaN HEMTs is 2D electron gas channel formation. The sheet carrier density and the confinement of the 2D electron gas located close to the interface of undoped and doped AlGaN/GaN heterostructures are due to the bending of the bands. The 2D electron gas is confined in a quantum well along the heterojunction and relies both on piezoelectric and spontaneous polarization induced effects. The piezoelectric effects can exert a substantial influence on the concentration and distribution of free carriers in strained group-III nitride heterostructures. Indeed, in AlGaN/GaN-based transistor structures, the piezoelectric polarization of the strained AlGaN barrier layer is more than five times that of AlGaAs/GaAs structures, which corresponds to an increasing current density. The very high mobility of confined electrons in the quantum well and high saturation velocity associated with GaN make up the key feature of AlGaN/GaN HEMTs.

The performance of AlGaN/GaN transistors is directly related to the electrical characteristics of the 2D electron gas formed at the interface. Since stress is a major factor in the operation and performance of AlGaN/GaN HEMTs, a thorough understanding of the impact of stress on performance can lead to improvements in device design. Al content and/or thickness of the AlGaN layer modify the carrier density and mobility in the 2DEG. For high-power microwave systems, AlGaN/GaN HEMT devices offer very high power densities. Thus, one aims at maximizing the product of the carrier density and the electron mobility in the 2DEG. It is now known that both the piezoelectric and the spontaneous polarization fields are dependent on the stress at the interface. To improve the properties of the 2D electron gas and reduce cracking of the AlGaN surface, strain engineering of the heterostructure is necessary. Strain engineered GaN-based HEMTs have been reported using AlN interlayers in the GaN bulk layer [30, 31].

7.6.2 Effect of Nitride Layer on Electrical Performance

In this case study, in a 2D GaN/AlGaN HEMT, the stress distribution resulting from intrinsic stress due to nitride passivation is studied using the VictoryProcess tool [32] from Silvaco. To calculate the stresses and evolutions of device geometry after each processing step, the so-called stress evolution" or stress history" model was used. We also show the presence of stress polarization as a consequence of nitride intrinsic stress besides spontaneous and piezoelectric polarization. The device structure generated from process simulation is used in the VictoryDevice tool [33] for device simulation of the output (I_d-V_g) characteristics. The effects of nitride intrinsic stress, which can be either compressive or tensile, on the electrical characteristics are studied in detail. The results are also compared for different tensor scale factors as well as different nitride layer thicknesses. We have investigated the role of spontaneous and piezoelectric polarization in nitride heterostructures, with particular emphasis on the design, characterization, and analysis of GaN/AlGaN HEMTs. A detailed understanding of polarization effects is essential in nitride heterostructure materials and device engineering.

The mechanism of the formation of 2DEG at the AlGaN/GaN interface has been discussed earlier. Consider a layer of AlGaN over a GaN layer without stress. They are both of gallium polarity. The lattice parameters of the AlGaN layer are lower than those of the GaN layer. The AlGaN layer is therefore stressed in tension. The polarization vectors in these two layers and the surface charges are shown [34].

This positive charge is compensated by the presence of electrons at the interface forming a gas confined to this interface which is called 2D electron gas. The AlGaN layer is called the "barrier," the layer GaN where the electrons are at the interface is called the "channel." Note that there is no intentional doping of the AlGaN barrier. Electrons confined to the interface can come from several sources: intrinsic faults, impurities in the GaN buffer layer, the AlGaN barrier, or surface conditions. Epitaxially grown III-V semiconductor materials with a wurtzite structure, such as GaN and AlN, exhibit a strong polarization effect. This polarization effect is critical in modeling nitride heterostructure or layered-structure devices, as it significantly alters the device performance [35].

Let us consider the case of the gallium polarity. Spontaneous polarization vectors do not change with stress. There are only the piezoelectric polarization vectors that are influenced by stress. We have three possibilities for the GaN layer strain on the substrate: without stress, the stress in tension, and stress in compression. Each case imposes a stress state for the AlGaN layer and thus a density of positive charge at the interface. The total polarization is the sum of spontaneous polarization and piezo-electric polarization. The presence of this polarization can have significant effects on device structures and has inspired us to study the effects of strain/stress applications for III-nitride devices.

In the first case, this is the optimum case because the stress in the AlGaN is not too large and the polarization vectors (spontaneous and piezoelectric) are in the same direction in the AlGaN. In the second case, the GaN layer is stressed in tension the piezoelectric polarization appears in GaN and increases the stress of AlGaN in tension. Theoretically, this case can generate a positive charge density, more raised to the interface than the previous case. However, the stress levels applied to the AlGaN would also be important causing the formation of dislocations or even cracks and consequently the material relaxation.

For GaN HEMT fabrication, the basic process steps are deposition, lithography, and etching. The basic process steps used are: initialization of sapphire (substrate), deposit AlN, deposit GaN, deposit AlGaN, deposit aluminum, etch aluminum, and deposit nitride. The basic structure after virtual fabrication is shown in Figure 7.10. To study the effects of nitride intrinsic stress on device performance, strain/stress distributions were simulated with the VictoryStress tool [36]. To model the intrinsic stress introduced into the nitride passivation layer during the deposition for the xx-component and the zz-component of the stress tensor, linear elasticity theory was used. The GaN crystal has orthotropic symmetry, meaning that its mechanical properties are isotropic in all directions within the basal plane of the wurtzite structure. The mechanical elastic constants for the c-axis (or growth direction), which is perpendicular to the basal plane are however different from those in the basal plane. General stress–strain relationship is used.

In epitaxial layers, the strain tensor can be represented by ε_{xx}, ε_{yy}, ε_{zz}, ε_{xy}, ε_{yz}, and ε_{zx}. The relationship between the various components of the strain tensor are given as follows:

$$\varepsilon_{xx} = \varepsilon_{yy} = \frac{a_S - a_0}{a_0} \qquad (7.25)$$

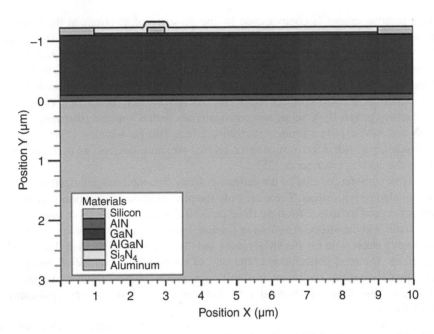

FIGURE 7.10   Basic GaN HEMT structure after the virtual fabrication was used in this study.

$$\varepsilon_{zz} = -2\frac{C_{13}}{C_{33}}\varepsilon_{xx} \qquad (7.26)$$

$$\varepsilon_{xy} = \varepsilon_{yz} = \varepsilon_{zx} = 0 \qquad (7.27)$$

where C_{13} and C_{33} are elastic constants. The stress distribution resulting from intrinsic stress in the nitride passivation layer is shown in Figure 7.11.

To study the field plate length vs. breakdown voltage characteristics, 2D simulations have been performed with the ATLAS device simulation tool for the heterostructure device analysis with field plates [37]. The transport models have been chosen in such a way that it should be computationally efficient with the highest precision. The hydrodynamic model has been implemented in this work because the Drift-Diffusion model is not suitable in terms of accuracy for submicron GaN devices [38]. The hole transport has been neglected because the AlGaN/GaN HEMTs are unipolar devices and a constant hole temperature of 300 K is maintained.

The basic semiconductor equations used for the simulation of HEMT are given below. The electron and hole current densities are given by the expressions [38]:

$$J_n = q \cdot \mu_n \cdot n \cdot \left(grad\left(\frac{\mathcal{E}_C}{q} - \psi \right) + \frac{k_B}{q} \cdot \frac{N_{C,0}}{n} \cdot rad\left(\frac{n \cdot T_L}{N_{C,0}} \right) \right) \qquad (7.28)$$

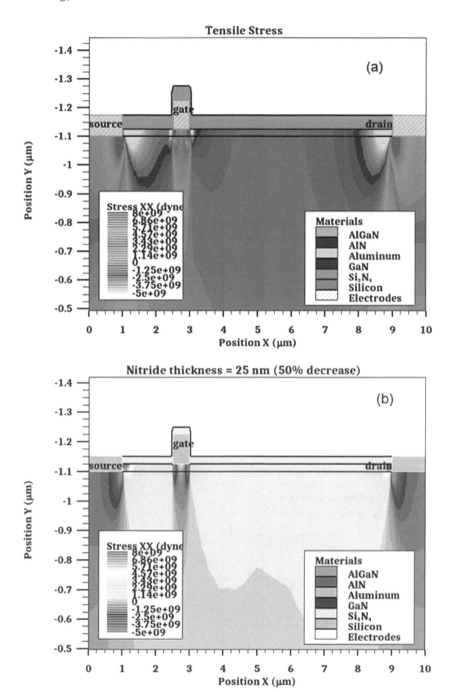

FIGURE 7.11 (a) An intrinsic (tensile) stress has been introduced by the nitride passivation layer during deposition. The *xx*-component of the stress is shown and (b) nitride layer thickness dependence (for 25 nm thick) is shown. The default unit of stress is dyne/cm².

$$J_p = q \cdot \mu_p \cdot p \cdot \left(grad\left(\frac{\mathcal{E}_V}{q} - \psi\right) - \frac{k_B}{q} \cdot \frac{N_{V,0}}{p} \cdot rad\left(\frac{p \cdot T_L}{N_{V,0}}\right) \right) \qquad (7.29)$$

where J_n is the electron current density,
    J_p is the hole current density,
    μ_n and μ_p are the carrier mobilities,
    $\mathcal{E}_C$ and $\mathcal{E}_V$ are the position-dependent band edge energies,
    $N_{C,0}$ and $N_{V,0}$ are the effective density of states, and
    T_L is the lattice temperature.

For the electron and hole mobilities, μ_n and μ_p respectively, the model considers a field dependence for the DD model and a carrier temperature dependence for the HD model. The energy transport equations for the HD model are given as:

$$div\,S_n = grad\left(\frac{\mathcal{E}_C}{q} - \psi\right) \cdot J_n - \frac{3 \cdot k_B}{2} \cdot \left(\frac{\partial(n \cdot T_n)}{\partial t} + R \cdot T_n + n \cdot \frac{T_n - T_L}{\tau_{\epsilon,n}}\right) \qquad (7.30)$$

$$div\,S_p = grad\left(\frac{\mathcal{E}_V}{q} - \psi\right) \cdot J_p - \frac{3 \cdot k_B}{2} \cdot \left(\frac{\partial(p \cdot T_p)}{\partial t} + R \cdot T_p + p \cdot \frac{T_p - T_L}{\tau_{\epsilon,p}}\right) \qquad (7.31)$$

where S_n and S_p are the energy fluxes.
    $\tau_{\epsilon,n}$ and $\tau_{\epsilon,p}$ are the energy relaxation times

$$S_n = -k_n \cdot grad\,T_n - \frac{5}{2} \cdot \frac{k_B T_n}{q} \cdot J_n \qquad (7.32)$$

where k_n is the thermal conductivity

$$k_n = \left(\frac{5}{2} + c_n\right)\frac{k_B^2}{q} \cdot T_n \cdot \mu_n \cdot n \qquad (7.33)$$

The DD mobility is modeled by

$$\mu_v^{LIF}(F_v) = \frac{\mu_v^{LI}}{\left(1 + \left(\frac{\mu_v^{LI} \cdot F_v}{v_v^{sat}}\right)^{\beta_v}\right)^{1/\beta_v}}, v = n, p \qquad (7.34)$$

where F_v represents the driving force for electrons and holes
    μ_v^{LI} is the zero-field mobility
    v_v^{sat} is the saturation velocity
    The HD mobility is modeled carrier temperature-dependent

$$\mu_v^{LIT} = \frac{\mu_v^{LI}}{1 + \alpha_v \cdot \left(T_v - T_L\right)} \tag{7.35}$$

where

$$\alpha_v = \frac{3 \cdot k_B \cdot \mu_v^{LI}}{2 \cdot q \cdot \tau_\epsilon \cdot \left(v_v^{sat}\right)^2} \tag{7.36}$$

And τ_ϵ is the energy relaxation times
v_v^{sat} is the saturation velocity.

To study the effects of intrinsic stress on the electrical performance, the I_d-V_g characteristics at a drain bias of 1 V are ramped from −6 to 1.0 V are shown in Figure 7.12 for different simulation conditions. The polarization parameters used in the simulation are shown in Table 7.6. Finite element modeling (FEM) techniques were applied to quantify the mechanical stress distribution in planar AlGaN/GaN structures.

The piezoelectric charges due to induced stress were scaled independently from those due to lattice mismatch stress and spontaneous polarization. The effects of nitride intrinsic stress, which can be either compressive or tensile, is visible on I_d-V_g characteristics. The results are also compared for different scale factors as well as

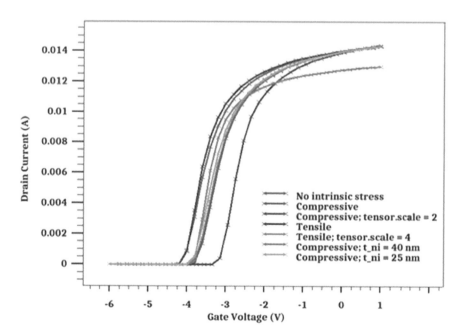

FIGURE 7.12 The I_d-V_g characteristics at a drain bias of 1 V when ramped from −6 V to 1.0 V for different stress conditions. Piezoelectric charges due to stress have been scaled.

TABLE 7.6
GaN Material Parameters Used in the Simulation

Parameter	GaN	Units
Bandgap	3.43	eV
Lattice Constant	3.189	A
Critical Elec. Field (E_c)	3.75	MV/cm
Hole Life Time (auger)	1×10^{-9}	s
Electron Life Time (auger)	1×10^{-9}	s
Peak Velocity	2.2×10^7	cm/s
Saturation Velocity	1.125×10^7	cm/s
Hole Mobility	10	cm²/Vs
Electron Mobility	525	cm²/Vs
Donor Energy Level	0.016	eV
Acceptor Energy Level	0.175	eV

different nitride layer thicknesses. At low to moderate Al concentrations, the 2DEG carrier concentrations observed are in very good agreement with those expected to arise from the combined effects of spontaneous and piezoelectric polarization [39]. It is observed that changes in mechanical stress result in a change in the charge density which in turn affects the maximum current in AlGaN/GaN transistors.

Using physics-based predictive TCAD simulations, we have studied the stress/strain distribution in AlGaN/GaN structures. The effects of mechanical stress and geometry dependence of the electrical characteristics are reported. Critical design issues involving spontaneous polarization, which does not change with stress, and the stress-dependent piezoelectric polarization are discussed.

7.6.3 FIELD PLATE DEPENDENT ELECTROSTATICS AND STRAIN MAPPING

Strain distributions within nanoscale non-planar device structures can be expected to behave differently compared to that of a planar structure due to the nature of the 3D device architecture. Because the modulation of strain becomes more pronounced with device scaling, the influence of the variation in strain on the device characteristics is of great concern. Therefore, as the geometries of these 3D device structures become more complex, it is important to understand how strain develops within the nanoscale HEMT device structures. In this case study, we perform simulations of a planar AlGaN/GaN and AlGaAs/GaAs HEMTs to investigate the effects of stress–strain and their mapping using the TCAD simulation tool. FEM of the mechanical stress in HEMT devices was done using VictoryDevice [33].

In the quest for alternative solutions for Si CMOS, III-V MOSFETs are being proposed. Due to their outstanding transport properties, III-V semiconductors hold a promise of reduced CMOS supply voltage without compromising the performance. Semiconductors composed of materials from columns III and V of the periodic table are promising candidates for use as the channel in MOSFET transistors (for both types of carriers). Due to their outstanding electronic properties, III-V materials such as GaAs, InAs, and $In_{1-x}Ga_xAs$ alloys have been widely used in the fields of optical communication, instrumentation, and detection. The basic building block now in

III-V technology for high frequencies is the HEMT. Nevertheless, the HEMTs will soon reach their limits in terms of the scaling laws, such as gate length and thickness of the heterostructures. Today, the III-V MOSFET technology is a very active area of research and has recently been included in the IRDS roadmap [40]. To benefit from the excellent intrinsic performance of III-V materials, several critical issues need to be resolved.

GaN has a hexagonal structure named "wurtzite" and has a bandgap energy of 3.4 eV. The energy bandgap, breakdown field, and electron mobility in GaN are very high which makes it extremely suitable to be used in power electronic applications [41]. For GaN-based HEMTs, one of the techniques to improve performance is the introduction of a field plate (FP) in the device [42]. The role of an FP is to modify the electric field profile and to decrease its peak value on the drain side of the gate edge, hence reducing high-field trapping effects and increasing breakdown voltage. Multiple field plates may be used to achieve both high breakdown voltage and high-frequency operation. Due to the piezoelectric nature of GaN, the 2DEG in AlGaN/GaN HEMT could be engineered by strain [43]. AlGaN/GaN HEMT has been a subject of intense investigation and have emerged as attractive candidates for high voltage, high-power operation at microwave frequencies. It is known that when the device is at pinch-off, the maximum electric field occurs at the drain side edge of the gate [44, 45]. FP technique is the most widely used approach and even employed in the commercial of AlGaN/GaN HEMTs for electric field (E-field) modulation and breakdown voltage improvements [46]. It has been reported that the gate breakdown occurs at the drain-side edge of that gate electrode due to the high electric field peak, via avalanche breakdown and thermally assisted tunneling [42].

The field plate and its extension are favorable to decrease the electric field intensity at the gate edge and hence reduce the electron trapping probability [47]. Understanding material and process limits for high breakdown voltage AlGaN/GaN HEMTs have been reported [48]. Due to the piezoelectric nature of GaN, the 2D electron gas in AlGaN/GaN HEMT can also be engineered by strain. Because the modulation of strain becomes more pronounced with device scaling, the influence of the variation in strain on the device characteristics is of great concern. Mechanical stress is an important factor influencing the performance of GaN-based devices. The material/electronic properties of compound semiconductors are dependent on stress. In highly piezoelectric materials like AlGaN/GaN, mechanical stress directly influences the piezoelectric polarization, and hence the charge density of the 2D electron gas [49]. Since stress is a major factor in the operation and performance of AlGaN/GaN HEMT devices, a thorough understanding of the impact of stress on performance can lead to improvements in device design.

Power devices are gaining more popularity nowadays with the extensive use of electronic devices in every field as these have a major influence on the system cost and efficiency. The importance of TCAD tools is also well known for the devices early-stage design to perform performance evaluation by using computer-based models to compensate for the expensive prototyping and large-signal experimental characterization of the millimeter and sub-millimeter wave devices. A combination of both the high transconductance and current gain cutoff frequency is desirable for power semiconductor devices, which is a major part of the power electronic systems.

GaN/SiC technology is preferred for industrial use as the SiC substrates are more cost-effective. The GaN/SiC HEMT technology can completely replace the presently used substrate/GaN epi-wafer technology due to the high thermal conductivity (+30%) of the semi-insulating SiC substrate along with its low cost. GaN has a hexagonal structure named "wurtzite" and has a bandgap energy of 3.4 eV. The energy bandgap, breakdown field, and electron mobility in GaN are very high which makes it extremely suitable to be used in power electronic applications [41]. The switching frequencies of GaN HEMTs can be very high in the range of GHz and the device shows high power conversion efficiency.

Device performance can be improved even without any change in the semiconductor material properties, however, with dedicated new device structures and fabrication methods. For example, for GaN-based HEMTs, one of the techniques to improve performance is the introduction of a field plate (FP) in the device [42]. The role of an FP is to modify the electric field profile and to decrease its peak value on the drain side of the gate edge, hence reducing high-field trapping effects and increasing breakdown voltage. The field plate technique has resulted in 2–4 times performance enhancement in RF GaN-based devices. Currently, the FP technique is the most widely used approach and even employed in the commercial of AlGaN/GaN HEMTs for electric field (E-field) modulation and breakdown voltage improvements [46]. The field plate and its extension are favorable to decrease the electric field intensity at the gate edge and hence reduces the electron trapping probability which results in the reduction of low-frequency noise [47]. Due to the piezoelectric nature of GaN, the 2D electron gas in AlGaN/GaN HEMT can also be engineered by strain. Strain distributions within nanoscale non-planar device structures can be expected to behave differently compared to that of a planar structure due to the nature of the 3D device architecture. Because the modulation of strain becomes more pronounced with device scaling, the influence of the variation in strain on the device characteristics is of great concern. Therefore, as the geometries of these 3D device structures become more complex, it is important to understand how strain develops within the nanoscale device structure. To investigate the field plate length dependence of the induced stress underneath the gate region, the stress distributions have been simulated using the TCAD simulation tool [50]. All the models used in the simulation, including the quantum models, were calibrated with previously reported results [42]. The detailed GaN material parameters and device geometry detail considered in the simulation have been listed in Table 7.6 and Table 7.7, respectively. The model and material parameters have also been incorporated in the 2D device simulator ATLAS [51].

7.6.4 MECHANICAL STRAIN/STRESS SIMULATION

Figure 7.13 shows the schematic cross-section view of the GaN HEMT with a field plate under simulation. Qualitative simulations were done using Silvaco ATLAS device simulation software to study the electrostatics in the device structure. The electrical field profile and the potential contours are shown in Figures 7.14 and 7.15, respectively. These simulations confirm that the potential contours and the electric field profile can be engineered by changing the lateral shift and the vertical

TABLE 7.7
Parameters for AlGaN/GaN HEMTs Used in the Simulation

Parameters	Value
Gate Length (L_g)	250 nm
Field Plate Length	0.3 μm
Cap and Spacer charge	-4×10^{14} cm^{-2}
Barrier and Cap charge	-2.5×10^{14} cm^{-2}
Semiconductor and Barrier charge	9.4×10^{14} cm^{-2}
Barrier Layer	50 nm
Cap Layer	5 nm

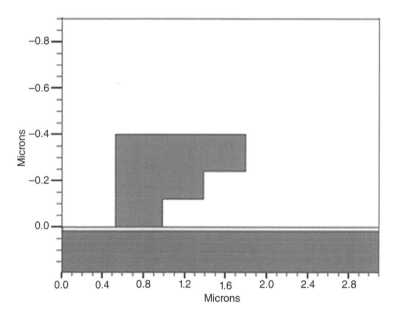

FIGURE 7.13   Schematic cross-section view of the GaN HEMT with a field plate.

height of the field plates. The role of an FP is to modify the electric field profile and to decrease its peak value on the drain side of the gate edge, which is visible from Figure 7.14.

Figure 7.14 shows the simulation of the E-field strength at a lateral cross section in the AlGaN region of the device in the off state. The first point to observe is that the E-field lines terminating at the field plates increase in magnitude with increasing drain bias. Figure 7.14 shows the electric field strength of a device with only the gate for an applied drain bias of 100 V. The lateral shift of the field plates and vertical thickness of the dielectric beneath them can be optimized to obtain E-field peaks of equal magnitude, each smaller than the critical field, to maximize the permissible drain bias of the device.

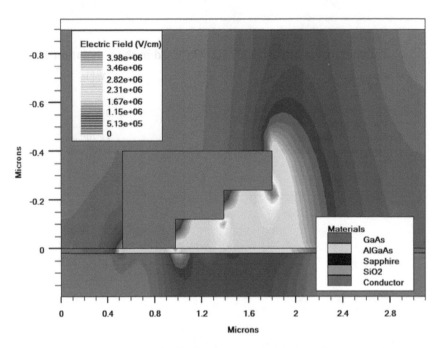

FIGURE 7.14 Simulated electrical field contours of a device with two field plates and applied drain bias of 100 V.

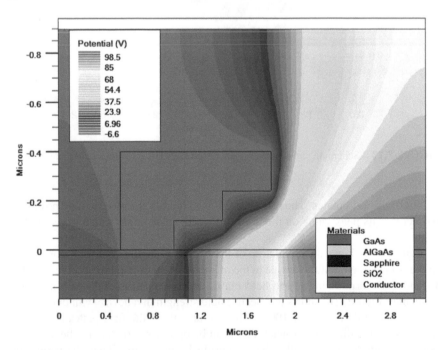

FIGURE 7.15 Simulated potential contours of a device with two field plates and applied drain bias of 100 V.

Mechanical stress represents a force per unit area. It is expressed in Pa or N/m². By convention tensile stress is represented as positive while compressive stress is negative. To represent the general state of stress at a given point of a material, one uses tensorial formalism. Under the effect of mechanical stress, a material is deformed. This deformation is not just in the direction of the stress. Also, the deformation depends on the mechanical properties of the material. It is assumed that the stress state of the material considered is in the elastic domain, that is, the stresses are sufficiently low not to generate plastic deformations, irreversible. This requires that the strain will be less than the yield strength of the material. Also, we will assume that the deformations are proportional to the strain, that is, the quantitative determination of the state of deformation of the material can be obtained using the formalism of linear elasticity theory. The stress–strain relation and also the relationship between the out-of-plane lattice parameter (c-parameter) to that of the in-plane lattice parameter (a-parameter) can be found from Hooke's law as discussed above.

The critical variables associated with the FP and the crucial semiconductor region beneath are readily seen to be the three geometrical variables, namely the FP length, the insulator thickness, and gate–drain separation, and the two material variables, namely the channel electron concentration over and the insulator dielectric constant. Stress simulation has been performed using the VictoryStress tool. Figure 7.16 shows the schematic cross-section view of the GaN HEMT with a field plate. To improve the off-state behavior, a single field plated structure is considered and the field plate is connected to the gate metal with a specific distance to the channel. The mechanical stress distribution due to the presence of the field plate is shown; (a) the Von Mises stress (Figure 7.16), (b) stress in the XX-direction (Figure 7.17), and (c) strain in the XX-direction (Figure 7.18). Stress generated is found to be compressive.

7.6.5 SUMMARY

Gallium nitride (GaN) is found to be an interesting material from the perspective of making a HEMT transistor delivering high power at high frequency. The potential of this material was approached by its intrinsic properties in comparison with other conventional semiconductors. The spontaneous and piezoelectric polarization, generated by the wurtzite structure, allows the creation of a 2D gas of electrons well confined to the interface where high electron density and mobility can be achieved. The simulation results showed that the strain provided by the passivation layer can induce a significant amount of piezoelectric charges in the submicron gate region. Therefore, strain engineering is believed to be an effective approach to adjust the threshold voltage of an AlGaN/GaN HEMT with a scaled gate length.

We have described in this chapter, the models showing the origin of electrons in 2D gas and the development of equations for the surface charge density. Also, we carried out a comparison of the properties of the different types of substrate. The substrate silicon appears to be the right choice for the realization of low-cost GaN-based HEMTs. The limits of conventional AlGaN/GaN HEMTs in terms of power performance and frequency have led to the emergence of InAlN/GaN technology, which has better potentials than its AlGaN/GaN counterpart. Given the current state of the art, the industry InAlN/GaN is promising, thanks to its advantages, in

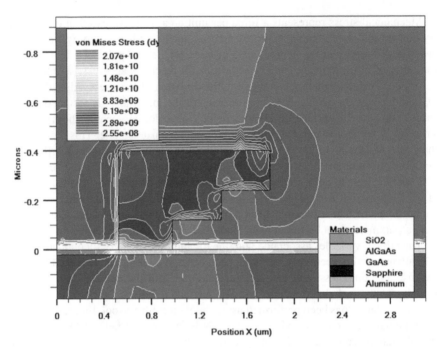

FIGURE 7.16   Von Mises stress distribution in the field plate region of the AlGaAs/GaAs HEMT structure.

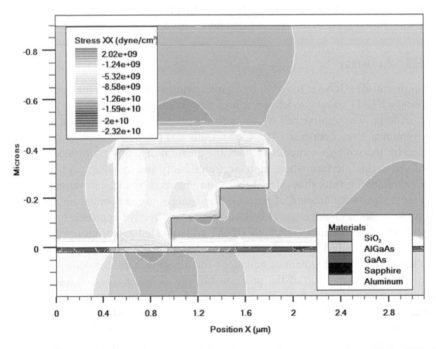

FIGURE 7.17   XX-stress distribution in the field plate region of the AlGaAs/GaAs HEMT structure.

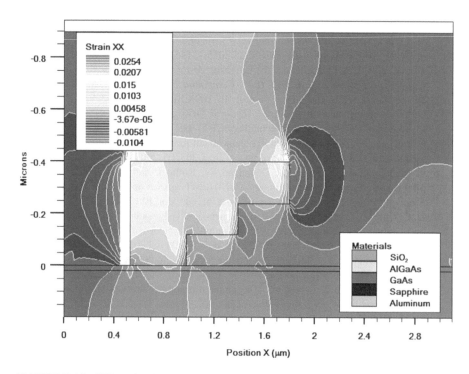

FIGURE 7.18 XX-strain distribution in the field plate region of the AlGaAs/GaAs HEMT structure.

particular, the high-density 2D gas carriers which lead to efficient operation at high frequencies. However, HEMT devices still exhibit dispersive effects that are not fully controlled and identified. Today, the analysis of the physical mechanisms responsible for the degradation of the performance of transistors is a necessary step for their reliability to validate the manufacturing technology used.

This work has contributed to the overall understanding of AlGaN/GaN HEMTs, and several techniques developed for the design of high breakdown voltage HEMTs which could potentially be beneficial in GaN technology. We have studied the electrostatics of the breakdown fields and map the strain/stress profile as a function of field plate length and height. The gate-connected field plates are very effective and by this technique, devices exhibited significant improvements. We have demonstrated that TCAD can be advantageously used to map stress/strain at the nanoscale. We propose TCAD as an alternative for strain analysis in advanced devices like GaN HEMTs.

REFERENCES

[1] M. Rais-Zadeh et al., "Gallium nitride as an electromechanical material," *J. Microelectromechanical Syst.*, vol. 23, no. 6, pp. 1252–1271, 2014, doi: 10.1109/JMEMS.2014.2352617.

[2] T. E. Kazior et al., *"More than Moore - wafer scale integration of dissimilar materials on a Si platform,"* in *2015 IEEE Compound Semiconductor Integrated Circuit Symposium (CSICS)*, 2015, p. 1, doi: 10.1109/CSICS.2015.7314517.

[3] T. E. Kazior, J. R. LaRoche, and W. E. Hoke, *"More than Moore: GaN HEMTs and Si CMOS get it together,"* in *2013 IEEE Compound Semiconductor Integrated Circuit Symposium (CSICS)*, 2013, pp. 1–4, doi: 10.1109/CSICS.2013.6659239.

[4] G. A. Armstrong and C. K. Maiti, *TCAD for Si, SiGe, and GaAs Integrated Circuits*. The Institution of Engineering and Technology (IET), UK, 2008.

[5] C. Y. Chang et al., "Development of enhancement mode AlN/GaN high electron mobility transistors," *Appl. Phys. Lett.*, vol. 94, no. 26, p. 263505, 2009, doi: 10.1063/1.3168648.

[6] F. Medjdoub, M. Zegaoui, D. Ducatteau, N. Rolland, and P. A. Rolland, "High-performance low-leakage-current AlN/GaN HEMTs grown on silicon substrate," *IEEE Electron Device Lett.*, vol. 32, no. 7, pp. 874–876, 2011, doi: 10.1109/LED.2011.2138674.

[7] D. C. Dumka, T. M. Chou, F. Faili, D. Francis, and F. Ejeckam, "AlGaN/GaN HEMTs on diamond substrate with over 7W/mm output power density at 10 GHz," *Electron. Lett.*, vol. 49, no. 20, pp. 1298–1299, 2013, doi: 10.1049/el.2013.1973.

[8] W. C. Johnson et al. "Nitrogen compounds of gallium 111. Gallic nitride," *J. Phys. Chem.*, vol. 682, no. 1912, pp. 2651–2654, 1930, doi: 10.1021/j150340a015.

[9] H. P. Maruska and J. J. Tietjen, "The preparation and properties of vapor-deposited single-crystal-line GaN," *Appl. Phys. Lett.*, vol. 15, no. 10, pp. 327–329, 1969, doi: 10.1063/1.1652845.

[10] H. Amano, N. Sawaki, I. Akasaki, and Y. Toyoda, "Metalorganic vapor phase epitaxial growth of a high quality GaN film using an AlN buffer layer," *Appl. Phys. Lett.*, vol. 48, no. 5, pp. 353–355, 1986, doi: 10.1063/1.96549.

[11] H. Amano, M. Kito, K. Hiramatsu, and I. Akasaki, "P-type conduction in Mg-doped GaN treated with low-energy electron beam irradiation (LEEBI)," *Jpn. J. Appl. Phys.*, vol. 28, no. 12 A, pp. L2112–L2114, 1989, doi: 10.1143/JJAP.28.L2112.

[12] M. A. Khan, J. M. Van Hove, J. N. Kuznia, and D. T. Olson, "High electron mobility GaN/AlxGa1-xN heterostructures grown by low-pressure metalorganic chemical vapor deposition," *Appl. Phys. Lett.*, vol. 58, no. 21, pp. 2408–2410, 1991, doi: 10.1063/1.104886.

[13] M. A. Khan, A. Bhattarai, J. N. Kuznia, and D. T. Olson, "High electron mobility transistor based on a GaN-AlxGa 1-xN heterojunction," *Appl. Phys. Lett.*, vol. 63, no. 9, pp. 1214–1215, 1993, doi: 10.1063/1.109775.

[14] S. Nakamura, T. Mukai, and M. Senoh, "Candela-class high-brightness InGaN/AlGaN double-heterostructure blue-light-emitting diodes," *Appl. Phys. Lett.*, vol. 64, no. 13, pp. 1687–1689, 1994, doi: 10.1063/1.111832.

[15] S. Nakamura et al., "InGaN-based multi-quantum-well-structure laser diodes," *Jpn. J. Appl. Phys.*, vol. 35, no. Part 2, pp. L74–L76, 1996, doi: 10.1143/jjap.35.l74.

[16] I. Vurgaftman and J. R. Meyer, "Band parameters for nitrogen-containing semiconductors," *J. Appl. Phys.*, vol. 94, no. 6, pp. 3675–3696, 2003, doi: 10.1063/1.1600519.

[17] J. Wu, "When group-III nitrides go infrared: New properties and perspectives," *J. Appl. Phys.*, vol. 106, no. 1, p. 11101, 2009, doi: 10.1063/1.3155798.

[18] G. Orsal et al., "Bandgap energy bowing parameter of strained and relaxed InGaN layers," *Opt. Mater. Express*, vol. 4, no. 5, pp. 1030–1041, 2014, doi: 10.1364/OME.4.001030.

[19] E. Sakalauskas et al., "Dielectric function and optical properties of Al-rich AlInN alloys pseudomorphically grown on GaN," *J. Phys. D. Appl. Phys.*, vol. 43, no. 36, Sep. 2010, doi: 10.1088/0022-3727/43/36/365102.

[20] E. Sakalauskas et al., "Dielectric function and optical properties of Al-rich AlInN alloys pseudomorphically grown on GaN," *J. Phys. D. Appl. Phys.*, vol. 43, no. 36, p. 365102, 2014, doi: 10.1088/0022-3727/43/36/365102.

[21] E. Iliopoulos, A. Adikimenakis, C. Giesen, M. Heuken, and A. Georgakilas, "Energy bandgap bowing of InAlN alloys studied by spectroscopic ellipsometry," *Appl. Phys. Lett.*, vol. 92, no. 19, p. 191907, 2008, doi: 10.1063/1.2921783.

[22] A. F. Wright, "Elastic properties of zinc-blende and wurtzite AlN, GaN, and InN," *J. Appl. Phys.*, vol. 82, no. 6, pp. 2833–2839, 1997, doi: 10.1063/1.366114.

[23] O. Ambacher et al., "Pyroelectric properties of Al(In)GaN/GaN hetero- and quantum well structures," *J. Phys. Condens. Matter*, vol. 14, no. 13, pp. 3399–3434, 2002, doi: 10.1088/0953-8984/14/13/302.

[24] J. A. Garrido, J. L. Sánchez-Rojas, A. Jiménez, E. Muñoz, F. Omnes, and P. Gibart, "Polarization fields determination in AlGaN/GaN heterostructure field-effect transistors from charge control analysis," *Appl. Phys. Lett.*, vol. 75, no. 16, pp. 2407–2409, 1999, doi: 10.1063/1.125029.

[25] A. T. Ping, Q. Chen, J. W. Yang, M. A. Khan, and I. Adesida, "DC and microwave performance of high-current AlGaN/GaN heterostructure field effect transistors grown on p-type SiC substrates," *IEEE Electron Device Lett.*, vol. 19, no. 2, pp. 54–56, 1998, doi: 10.1109/55.658603.

[26] J. P. Ibbetson, P. T. Fini, K. D. Ness, S. P. DenBaars, J. S. Speck, and U. K. Mishra, "Polarization effects, surface states, and the source of electrons in AlGaN/GaN heterostructure field effect transistors," *Appl. Phys. Lett.*, vol. 77, no. 2, pp. 250–252, 2000, doi: 10.1063/1.126940.

[27] J. K. Sheu, M. L. Lee, and W. C. Lai, "Effect of low-temperature-grown GaN cap layer on reduced leakage current of GaN Schottky diodes," *Appl. Phys. Lett.*, vol. 86, no. 5, pp. 1–3, 2005, doi: 10.1063/1.1861113.

[28] A. Fischer, H. Kühne, and H. Richter, "New approach in equilibrium theory for strained layer relaxation," *Phys. Rev. Lett.*, vol. 73, no. 20, pp. 2712–2715, 1994, doi: 10.1103/PhysRevLett.73.2712.

[29] J. Xie, X. Ni, M. Wu, J. H. Leach, Ü. Özgür, and H. Morkɔ, "High electron mobility in nearly lattice-matched AlInNAlNGaN heterostructure field effect transistors," *Appl. Phys. Lett.*, vol. 91, no. 13, p. 132116, 2007, doi: 10.1063/1.2794419.

[30] S. Nakajima, "*GaN HEMTs for 5G base station applications*," in *2018 IEEE International Electron Devices Meeting (IEDM)*, 2018, pp. 14.2.1–14.2.4, doi: 10.1109/IEDM.2018.8614588.

[31] A. Nakajima, Y. Sumida, M. H. Dhyani, H. Kawai, and E. M. Narayanan, "GaN-based super heterojunction field effect transistors using the polarization junction concept," *IEEE Electron Device Lett.*, vol. 32, no. 4, pp. 542–544, 2011, doi: 10.1109/LED.2011.2105242.

[32] Silvaco International, VictoryProcess user manual, 2018.

[33] Silvaco International, VictoryDevice user manual, 2018.

[34] O. Ambacher et al., "Two-dimensional electron gases induced by spontaneous and piezoelectric polarization charges in N- And Ga-face AlGaN/GaN heterostructures," *J. Appl. Phys.*, vol. 85, no. 6, pp. 3222–3233, 1999, doi: 10.1063/1.369664.

[35] S. H. Park and S. L. Chuang, "Spontaneous polarization effects in wurtzite GaN/AlGaN quantum wells and comparison with experiment," *Appl. Phys. Lett.*, vol. 76, no. 15, pp. 1981–1983, 2000, doi: 10.1063/1.126229.

[36] Silvaco International, VictoryStress user manual, 2018.

[37] S. Vitanov et al., "*Predictive simulation of AlGaN/GaN HEMTs*," in *2007 IEEE Compound Semiconductor Integrated Circuits Symposium*, 2007, pp. 1–4, doi: 10.1109/CSICS07.2007.31.

[38] S. Vitanov et al., "Physics-based modeling of GaN HEMTs," *IEEE Trans. Electron Devices*, vol. 59, no. 3, pp. 685–693, 2012, doi: 10.1109/TED.2011.2179118.

[39] P. L. F. Sacconi, A. Di Carlo, and H. Morkoc, "Spontaneous and piezoelectric polarization effects on the output characteristics of AlGaN/GaN heterojunction modulation doped FETs," *IEEE Trans. Electron Devices*, vol. 48, no. 3, pp. 450–457.

[40] IEEE, "International roadmap for devices and systems, 2020 update, beyond CMOS." 2020.

[41] U. K. Mishra, P. Parikh, and Y. F. Wu, "AlGaN/GaN HEMTs - An overview of device operation and applications," *Proc. IEEE*, vol. 90, no. 6, pp. 1022–1031, 2002, doi: 10.1109/JPROC.2002.1021567.

[42] S. Karmalkar and U. K. Mishra, "Enhancement of breakdown voltage in AlGaN/GaN high electron mobility transistors using a field plate," *IEEE Trans. Electron Devices*, vol. 48, no. 8, pp. 1515–1521, 2001, doi: 10.1109/16.936500.

[43] W. C. Cheng et al., "Silicon nitride stress liner impacts on the electrical characteristics of AlGaN/GaN HEMTs," 2019, doi: 10.1109/EDSSC.2019.8754212.

[44] Z. Wang, H. Zhang, and J. B. Kuo, "Turn-OFF transient analysis of super junction IGBT," *IEEE Trans. Electron Devices*, vol. 66, no. 2, pp. 991–998, 2019.

[45] R. Vetury et al., "The impact of surface states on the DC and RF characteristics of AlGaN/GaN HFETs," *IEEE Trans. Electron Devices*, vol. 48, no. 3, pp. 560–566, 2001, doi: 10.1109/16.906451.

[46] Y. Ando, Y. Okamoto, H. Miyamoto, T. Nakayama, T. Inoue, and M. Kuzuhara, "10-W/ mm AlGaN-GaN HFET with a field modulating plate," *IEEE Electron Device Lett.*, vol. 24, no. 5, pp. 289–291, 2003, doi: 10.1109/LED.2003.812532.

[47] H. C. Chiu, C. W. Yang, H. C. Wang, F. H. Huang, H. L. Kao, and F. T. Chien, "Characteristics of algan/gan hemts with various field-plate and gate-to-drain extensions," *IEEE Trans. Electron Devices*, vol. 60, no. 11, pp. 3877–3882, 2013, doi: 10.1109/TED.2013.2281911.

[48] Y. Dora, A. Chakraborty, L. McCarthy, S. Keller, S. P. Denbaars, and U. K. Mishra, "High breakdown voltage achieved on AlGaN/GaN HEMTs with integrated slant field plates," *IEEE Electron Device Lett.*, vol. 27, no. 9, pp. 713–715, 2006, doi: 10.1109/ LED.2006.881020.

[49] S. Joglekar, C. Lian, R. Baskaran, Y. Zhang, T. Palacios, and A. Hanson, "Finite element analysis of fabrication- and operation-induced mechanical stress in AlGaN/GaN transistors," *IEEE Trans. Semicond. Manuf.*, vol. 29, no. 4, pp. 349–354, 2016, doi: 10.1109/ TSM.2016.2600593.

[50] Silvaco International, VictoryDevice user manual, 2019.

[51] Silvaco International, ATLAS device simulator, 2018.

8 Strain-Engineered SiGe Channel TFTs for Flexible Electronics

In more than 60 years, the semiconductor industry has grown exponentially, increasing the performance and density of integration of components while reducing manufacturing costs [1]. Anticipating a limit to this continuous miniaturization (More-Moore), intense research efforts have been invested to co-integrate various functionalities (More-than-Moore) [2]. In this context, flexible and even stretchable electronics offer many opportunities [3]. The organic and printed electronics industry makes it possible to manufacture at low cost and on large areas of flexible devices [4] but whose complexity and high-frequency characteristics are limited [5]. Organic semiconductors, and amorphous or polycrystalline Si, which can be processed at relatively low temperatures and with low cost, often suffice to address them.

The development of functional flexible electronics is essential to enable applications such as conformal medical imagers, wearable health monitoring systems, and flexible light-weight displays. Intensive research on thin-film transistors (TFTs) is being conducted to produce high-performance devices for improved backplane electronics. However, there are many challenges regarding the performance of devices fabricated at low temperatures that are compatible with flexible plastic substrates. Prior work has reported on the change in TFT characteristics due to mechanical strain, with especially extensive data on the effect of strain on field-effect mobility. We shall investigate the effects of gate-bias stress and elastic strain on the long-term stability of flexible low-temperature hydrogenated amorphous silicon (a-Si:H) TFTs, as the topic has yet to be explored systematically [6].

Researchers have demonstrated functional flexible TFTs fabricated from a variety of materials. The research documented in this thesis focuses on hydrogenated amorphous silicon (a-Si:H) TFTs, as they offer several key advantages. Firstly, a-Si:H TFTs are currently in widespread commercial production, most notably for application in at-screen display panels. Although organic TFTs may be simple and cheap to fabricate as well [7, 8], amorphous silicon has relatively high mobility compared to organic materials. Crystalline and polycrystalline silicon can offer higher mobility but also require high-temperature processes that would damage flexible plastic substrates or additional unconventional process steps. High mobility is necessary for good current driving ability, which allows a large amount of current to be controlled by a device with a small area. This is important for applications such as the drive TFT device size minimizes the area of the pixel circuit, as such, more space is then available for a larger area which results in a higher resolution display.

229

Currently, flexible electronics are mainly addressing electronic applications operating at low or moderate speed. Several flexible electronic applications require the operating speed (frequency) to be beyond 1 GHz and hence these applications can be further termed as radio frequency (RF) flexible electronics. RF flexible electronics are more attractive than traditional flexible electronics because of their versatile capabilities, dramatic power savings when operating at reduced speed, and a broader spectrum of applications.

Improvement of Si-based device speed implies significant technical and economic advantages. Transferrable single-crystalline Si nanomembranes (SiNMs) are preferred to other materials for flexible electronics owing to their unique advantages. While the mobility of bulk Si can be enhanced using strain techniques, implementing these techniques into transferrable single-crystalline SiNMs has been challenging [9]. A major challenge of developing flexible displays and sensors with amorphous silicon is electrical metastability, such as threshold voltage, which gradually shifts over time when a gate bias is applied. The metastability leads to a degradation in the operation of TFT circuits and limits the device lifetime. This phenomenon leads to a decrease in the brightness of the pixels and eventually causes the display to become perceptibly dimmer. To advance flexible a-Si:H TFT technology, it is necessary to gain a better understanding of the shift mechanisms under the influence of applied mechanical strain. Such knowledge could then be applied at the device level to design more robust flexible TFTs. Alternatively, systems can be designed to compensate for instability using knowledge of long-term device behavior. The effects of applied strain, occurring when a flexible device is bent, on the physical mechanisms of instability are not well characterized. Hence, it is the goal of the research presented in this thesis to investigate the impact of simultaneous electrical and mechanical stress on the behavior of flexible a-Si:H TFTs.

For flexible applications, form factors such as bendability and large areas are of more importance. There is a wider spectrum of electronics applications where higher speed and mechanical flexibility are simultaneously needed, such as high-speed and wireless communications, remote sensing, and airborne/space surveillance. The flexible nanomembranes are virtually dislocation-free and have many potential new applications. Fast flexible electronics provide superior performance and application advantages. As is known, high-speed devices consume much less power if they are operated at a reduced speed, which dramatically benefits battery-powered devices. Wirelessly connected devices enabled only by high operation speed/frequency are more convenient to use than wired devices and a high-frequency wireless system is also generally more compact than a low-frequency one. Strain engineering also provides opportunities for massively parallel self-assembly of a wide variety of 3D nanostructures. In the flexible electronics sector, there is an urgent need for higher performance, and hence many new technologies are emerging in recent years. The field of organic and printed electronics allows low production costs and large areas. The silicon technology allows nanometric resolutions and better performance.

Strain engineering is recognized as an effective technique to boost the performance of both n- and p-MOS transistors. Strain engineering of Si offers the potential to increase carrier mobility, and hence device performance [10], in applications spanning areas as diverse as nanoelectromechanical systems (NEMS), nanophotonics, and high-speed electronics. A new TCAD based design methodology is needed to combine both mechanical flexibility and high performance. This work presents a predictive technology CAD calibration methodology for SiGe-channel TFTs, employing physics-based modeling to support and enhance simulation-based device development. We demonstrate the combination of strained-SiGe applying a strain-sharing scheme between Si and SiGe multiple epitaxial layers, to create strained TFTs. A new speed record of Si/SiGe-based TFT for flexible electronics without using aggressively scaled critical device dimensions has been demonstrated.

In this chapter, we shall study the applications of SiGe heteroepitaxial nanostructures toward flexible devices. Based on the nonlinear finite element technique, a TCAD framework has been developed to introduce different types of large deformations and obtain the local stress/strain profiles (mapping) inside the devices. The impacts of various kinds of deformations on the electrical characteristics of the strained TFTs are studied. The feasibility (via predictions) for SiGe channel TFTs, combining considerable mechanical flexibility, high electrical performance, and finally, excellent stability under deformation may be useful for mechanical deformation-aware design and modeling for highly flexible electronics. This is the highest cutoff frequency f_T (>700 MHz) possible for strain-engineered SiGe TFTs as predicted in entirely flexible technologies that can be used to realize active transceivers operating in the MHz regime.

8.1 HETEROEPITAXY OF SI-GE LAYERS

Epitaxial growth technique has been one of the cornerstones of the semiconductor industry, which is used in almost all semiconductor devices, and in particular, the heteroepitaxy is ubiquitously used to fabricate hetero-interfaces and heterostructures which control flows of charge carriers in semiconductor devices. The epitaxial strain commonly called strain engineering is regarded as another degree of freedom for tuning physical properties of semiconductors including band gap and charge carrier mobility. It is well known that the heteroepitaxial growth is dramatically affected by the mismatch strain due to the different lattice constants between the substrate and deposited layer materials. It has attracted lots of research interests to rationalize the influence of strain on device characteristics [11].

Heteroepitaxy of silicon–germanium (SiGe) is commonly regarded as the stereotype of semiconductor epitaxy. While the SiGe material system has attracted a tremendous amount of attention due to its applications for band-gap engineering in the microelectronic industry, the major challenge facing the development of new SiGe-based devices remains the controllable epitaxial growth of self-assembled nanostructures. Even if the innovation in Si-based devices has been boosted recently by the development of ultra-thin body fully depleted silicon on insulator transistors, a real breakthrough would be the demonstration of flexible devices by group IV elements since it allows the convenient integration into the nowadays silicon microelectronics. An overall understanding of the behavior of the flexible device is still partially missing due to the complexity and the interplay of kinetics and energetic driving forces, preventing the development of new devices. For SiNMs, the fabrication approach involves epitaxial growth of a strained-SiGe thin film (the stressor layer) on an ultra-thin Si template layer of SOI, followed by an epitaxial Si layer on top of the SiGe and subsequent "release" from the handle substrate by chemical removal of the buried oxide layer.

This approach offers opportunities for an enormous variety of novel integration paradigms, such as in flexible electronics [12], where direct growth of strained-Si is not possible. Furthermore, 3D nanostructures can be designed by manipulating the strain sharing, such that the release process causes curling along with preselected directions, and spatially varying strain fields can be established in membranes via the use of "local stressors" [13]. There exists a considerable parameter space for engineering strain in nanomembranes, via the thickness of layers, the composition of layers, starting substrate, balancing or unbalancing of layer thicknesses, the introduction of dopant layers, the introduction of surface layers, and so on. For an ideal, elastically balanced trilayer structure (Si/SiGe/Si) with coherent interfaces, consideration of force balance between the layers allows the strain distribution between the alloy layer and the Si layers.

Germanium has the same crystalline structure as Silicon. However, Si and Ge have a lattice mismatch of around 4.17%, making possible $Si_{1-x}Ge_x$ alloys in the full range of germanium concentrations (i.e., $0 < x < 1$). A lattice mismatch of 4.2% between Si and Ge allows for a progressive increase in the virtual substrate's lattice constant, via multilayer compositional grading from pure Si through to the final $Si_{1-x}Ge_x$ alloy composition. This use of multiple low-mismatch interfaces encourages relaxation via dislocation propagation and minimizes dislocation nucleation resulting in a final relaxed SiGe layer with fewer defects than would be observed if a single high-mismatch alloy layer had been used. Due to the great compatibility of silicon and germanium, many efforts have been made over the last 40 years on SiGe deposition processes on the silicon substrate. Elastic strain sharing between thin heteroepitaxial Si and SiGe films, enabled by techniques that allow the release of these films from a handling substrate, creates a new material: freestanding, single-crystal, strained nanomembranes.

The concept, which would have implications far beyond the increased device speed in Si devices, particularly in the integration of new functions in all kinds of electronic devices, is based on the plastic deformation of a "compliant" layer during elastic strain relaxation of the crystalline layers it supports. This compliant layer distorts during strain sharing between the template layer (which rests on top of the compliant layer) and a pseudomorphic film is grown on the template layer, typically via a high-temperature process. Another remarkable property of SiGe materials is the possibility of being able to control the stress, naturally induced up to a certain film thickness (critical thickness). This allows forming structures with high mobility due to the modification of the band structures of the layers concerning the same relaxed materials.

For the bulk $Si_{1-x}Ge_x$ material, the lattice constant varies linearly with the content of germanium which is approximated by Vegard's law with a small deviation. This is given in Equation (8.1) with an accuracy of about 10^{-4} nm [14].

$$a_{Si_{1-x}Ge_x} = 0.5431 + 0.01992x + 0.0002733x^2 \ (nm) \tag{8.1}$$

The $Si_{1-x}Ge_x$ layer can be biaxially strained if a very thin layer is grown on a relaxed $Si_{1-y}Ge_y$ substrate. The strain in the top layer is compressive for $x > y$ (Figure 8.1b) and it is tensile for $x < y$ (Figure 8.1d). There is an in-plane component of the strain ($\epsilon_x = \epsilon_y = \epsilon_{\parallel}$) along with a perpendicular component ($\epsilon_\perp$) (shown in Figure 8.1b). Both the strain components are related according to isotropic elasticity theory as:

$$\epsilon_\perp = \frac{-2v}{1-v}\epsilon_{\parallel} \tag{8.2}$$

where v is the Poisson's ratio.

Due to the distortion in the lattice, a parallel lattice constant is produced (shown in Figure 8.1b). For the layers of thicknesses h_A and h_B having lattice parameters a_A and a_B respectively, where $a_A < a_B$, (Figure 8.1a), it is given as:

$$a_{\parallel} = a_A\left[1+\frac{f}{\left(1+\dfrac{G_A h_A}{G_B h_B}\right)}\right] \tag{8.3}$$

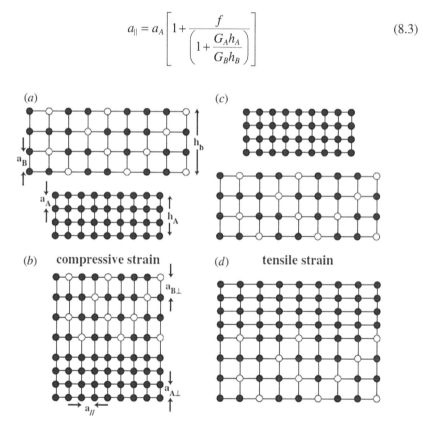

FIGURE 8.1 Schematic diagrams (a) of the lattice constant of thin $Si_{1-x}Ge_x$ layer (a_B) and bulk-silicon layer (a_A), (b) tetragonal lattice distortion when the thin $Si_{1-x}Ge_x$ layer (compressively strained) is grown on bulk-Si substrate, (c) lattice constants of the bulk-Si layer which has to be grown on a bulk-$Si_{1-y}Ge_y$ layer, and (d) tensile strain in the Si layer when it is grown on top of the $Si_{1-y}Ge_y$ layer [15].

where h_i are the thicknesses and G_i represents the shear moduli of each layer. The misfit between the two layers is denoted as f, which is defined as

$$f = \frac{a_B - a_A}{a_A} \tag{8.4}$$

The parallel or in-plane component of strain in layer A ($\epsilon_\parallel^A$) is defined by the expression $\dfrac{f}{\left(1 + \dfrac{G_A h_A}{G_B h_B}\right)}$, and its relationship with the corresponding in-plane strain in

layer B ($\epsilon_\parallel^B$) is given as:

$$\epsilon_\parallel^A = -\left(\frac{G_B h_B}{G_A h_A}\right)\epsilon_\parallel^B \tag{8.5}$$

When a layer of SiGe is grown by adapting its material parameter to that of the substrate and thus undergoing strain deformation, it is known as pseudomorphic growth. Such growth is, however, only possible if the thickness of the deposited film remains lower than a value h_c which is called the critical thickness. When the thickness of the film becomes greater than the critical thickness, the energy accumulated by the system becomes maximum and it tends to minimize its total energy by relaxation of the induced stress. Structural defects are thus likely to appear. The critical thickness, therefore, represents the maximum energy limit that the system can store during growth by adapting to the crystal lattice imposed by the substrate, before undergoing irreversible structural deformations.

The critical thickness of the strained layer can be predicted by the thermodynamic equilibrium model given by Van der Merwe [16]. At equilibrium, the total energy of the system is minimized, and also an array of dislocations are created. The critical thickness is given as:

$$h_c \approx \frac{19}{16\pi^2}\left(\frac{1+v}{1-v}\right)\left(\frac{b}{f}\right) \tag{8.6}$$

where b (slip distance) $= a_A/\sqrt{2}\ a_A$: substrate (relaxed) lattice constant.

8.1.1 ELECTRONIC PROPERTIES OF SIGE LAYERS

Bandgap engineering using Si/SiGe heterostructures has the potential to enhance device performance. The effect of the heterostructure material parameters on the band structures of the conduction and valence bands needs to be understood to obtain the carrier effective masses and transport properties. The band structure of compressively strained SiGe layers, as well as strain, relaxed virtual substrates have been calculated using pseudopotentials or k.p techniques [17].

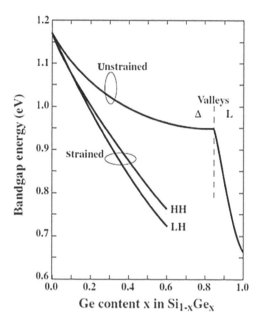

FIGURE 8.2 Bandgap of bulk (unstrained) $Si_{1-x}Ge_x$ and that of strained $Si_{1-x}Ge_x$ grown on bulk-silicon substrates [15].

The bandgap for unstrained (100) $Si_{1-x}Ge_x$ material is shown in Figure 8.2. The bandgap energy of $Si_{1-x}Ge_x$ layers decreases continuously with the germanium content, with an acceleration of decay for atomic concentrations of germanium exceeding 85%. This point corresponds to the change of the minimum conduction band from the Δ_6 valleys like silicon to the L valleys as in the case of germanium. We also observe the existence of two curves for compressively strained $Si_{1-x}Ge_x$ that can be explained by a lifting of the degeneracy of the maximum of the valence band (separation from the band of heavy holes and light holes). Finally, we can conclude that indirect band energy is decreasing with the compressive stress of the SiGe layers and the germanium content.

As the bandgap of $Si_{1-x}Ge_x$ is smaller than that of Si, a compressively strained-$Si_{1-x}Ge_x$ layer grown on silicon substrate results in a type I band discontinuity with a considerably larger valence band discontinuity than the conduction band discontinuity. This creates a quantum well for holes as shown in Figure 8.3a. This can be also achieved with a strained-$Si_{1-x}Ge_x$ layer grown on relaxed-$Si_{1-y}Ge_y$ substrates (where $x > y$). Similarly, electron confinement is possible for large conduction band discontinuity in case of tensile strained-Si or $Si_{1-x}Ge_x$ layer grown on the relaxed-$Si_{1-y}Ge_y$ virtual substrate where $x < y$ (shown in Figure 8.3b). The energy of the nth sub-band in the quantum well of width w according to the infinite barrier approximation is given by:

$$E_n = \frac{(n+1)^2 \pi^2 \hbar^2}{2m^* w^2} \text{ for n} = 0,1,2,3,\ldots\ldots \tag{8.7}$$

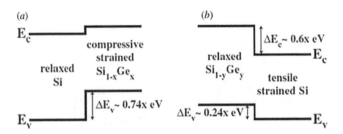

FIGURE 8.3 Band discontinuities for (a) strained-$Si_{1-x}Ge_x$ layer (compressive strain) grown on the relaxed silicon substrate and (b) strained-Si layer (tensile strain) grown on relaxed $Si_{1-y}Ge_y$.

The bandgap of strained-Si and strained-$Si_{1-x}Ge_x$ are smaller in comparison to that of the unstrained materials, which has been empirically given as [18]:

$$E_g^{Si}(x) = 1.11 - 0.4x \tag{8.8}$$

$$E_g^{SiGe}(x) = 1.11 - 0.74x \tag{8.9}$$

The conduction band discontinuity (ΔE_c) is found to be 160 meV [19], for a strained-Si layer on a $Si_{0.7}Ge_{0.3}$ substrate.

8.1.2 MOBILITY

In the technologically important case of Si (001), tensile strain splits the 6-fold degenerate conduction band minimum into a higher-energy 4-fold degenerate valley and a lower-energy 2-fold degenerate valley. The valence band is altered by lifting the degeneracy of the heavy and light hole (LH and HH) bands. These two effects suppress intervalley/band scattering and reduce the effective transport mass, resulting in significant carrier mobility enhancement. Improvements of 125% in electron mobility and 120% in hole mobility have been demonstrated in strained-Si metal-oxide-semiconductor field effect transistors.

The biaxial strain has a considerable effect on the Si and SiGe band structures which leads to an enhancement in carrier mobility. Due to the tetragonal strain, band splitting takes place which converts the 6-fold degeneracy of the conduction band into 2-fold and 4-fold degeneracies. More electrons are occupied in the Δ_2 valleys which have lower energy. As the Δ_2 valleys are associated with lower effective mass, the mobility of electrons increases due to reduced transverse in-plane effective mass (m_t) whereas the longitudinal out-of-plane mass (m_l) increases. Mobility is inversely related to effective mass as given by [20]:

$$\mu = \frac{q\tau}{m^*} \tag{8.10}$$

where m^* represents the conductivity effective mass and $1/\tau$ is the scattering rate. The mobility is enhanced under strain due to a reduction in the conductivity effective mass and also the scattering rate. The total electron mass, m^* is calculated by the summation of mobility components from the six degenerate valleys, which is given as [21]:

$$m^* = \left[\frac{1}{6} \left(\frac{2}{m_l} \right) + \left(\frac{4}{m_t} \right) \right]^{-1}$$

(8.11)

The effective mass in the direction perpendicular to the axis (transverse mass) is given as $m_t = 0.19m_o$ which is considerably smaller than that in the parallel direction (longitudinal mass) given by $m_l = 0.98m_o$, m_o being the free electron mass. The mass, as well as the change in scattering rates, are responsible for the improvement in mobility. It has been theoretically predicted that at approximately 0.8% strain (which results from a $Si_{0.8}Ge_{0.2}$ alloy), electrons are occupied only in the 2-fold sub valleys, and hence the effective mass decreases significantly. This also reduces the intervalley scattering that previously occurred because there are now fewer possible final states for a carrier to scatter into. In the case that only the 2-fold degenerate valleys are populated, the only intervalley scattering occurs in between them.

The strain has a similar effect on the valence band of silicon. The LH and HH bands are degenerate at the zone center. The labels of LH and HH bands arise because of the effective masses of carriers occupying these states. The third band is the split-off band, which is 44 meV higher in energy due to spin–orbit interactions. The addition of strain lifts the degeneracy of the LH and HH bands and causes the split-off band to be further removed from the other two, reducing further the possibility of any interband scattering events involving it. Along with this, the hole population is increased in the light hole band compared to the heavy hole band which results in a low in-plane conduction mass. The reduction in interband scattering and effective mass is responsible for higher hole mobility.

The mobility of the carriers is limited by various scattering phenomena like Coulomb, phonon, and surface roughness scattering. The phonon scattering dominates the carrier mobility at room temperature. However, at low temperatures, the scattering mechanisms responsible are impurities scattering, surface roughness scattering, alloy scattering, etc. Scattering in an alloy is due to the uneven distribution of the constituents which gives rise to fluctuations in the carrier potential. The scattering rate in $Si_{1-x}Ge_x$ alloy is a function of the scattering potential, V_a and Ge mole fraction, x. The expression for mobility-limited by alloy scattering is given as [22]

$$\mu_{alloy} = \frac{e\hbar^3}{m_t^{*2}\Omega_o V_a^2 x(1-x)} \left(\frac{16}{3b} \right)$$

(8.12)

where the transverse effective mass is m_t^*, Ω_o represents the atomic volume, and the variational parameter b is given by

$$b = \left(\frac{33m_l^* e^2 n}{8\epsilon\hbar^2} \right)^{1/3}$$

(8.13)

where n is the carrier concentration and m_1^* is the longitudinal effective mass. The alloy scattering limited mobility is assumed to be similar in relaxed and strained-SiGe materials. The value of alloy scattering potential, V_a used in the calculation is 0.7 eV to fit the experimental data. If the alloy potential value is 1.4 eV, the hole mobilities in the strained $Si_{1-x}Ge_x$ alloys based on Monte Carlo simulation decrease by a factor of 2–3 in comparison to the mobility using 0.7 eV up to the doping concentration of 10^{19} cm^{-3}.

The role of alloy scattering in limiting the mobility becomes less important for a doping concentration of $\geq 10^{19}$ cm^{-3} as the mobility does not vary much with Ge content. Taking these facts into account, the hole mobility enhancement in the strained-Si and SiGe layers has been modeled as [23]

$$\mu p, stSi = \mu_{Si} \cdot \left(1 + 4.31x - 2.28x^2\right) \tag{8.14}$$

The mobility limited by alloy scattering can be expressed as:

$$\left[\mu_{alloy}\right]^1 = x(1-x)\exp(-7.68x)/124.1 \text{ for } x \leq 0.2 \text{ and}$$
$$\left[\mu_{alloy}\right]^1 = \exp(-2.58x)/2,150 \text{ for } 0.2 < x < 0.6 \tag{8.15}$$

$$\left[\mu_{SiGe}\right]^{-1} = \left[\mu_{stsi}\right]^{-1} + \left[\mu_{alloy}\right]^{-1} \tag{8.16}$$

For 28% Ge content in the buffer layer which corresponds to ~1.2% strain, the peak hole mobility enhancement is 45% over that of Si, however, it decreases at the high effective electric field. Also, slight degradation in hole mobility is observed in comparison to the Si layer for lower levels of strain in the Si layer (13% Ge content in the SiGe buffer layer). While the compressive strain is useful only for the hole mobility enhancement, the tensile strain can improve electron as well as hole mobilities by reducing the carrier effective masses and suppressing intervalley phonon scattering. Under tensile strain (positive strain value), more electrons are occupied in the lower energy Δ_2 valleys and hence the in-plane electron mobility increases significantly. Reduction in the electron conductivity mass is mostly responsible for the mobility increment along with the small decrease of intervalley scattering. For nearly 0.8% tensile strain in the Si layer (grown over a fully relaxed $Si_{0.8}Ge_{0.2}$ buffer), the electron mobility reaches ~2,300 cm^2/Vs.

Instead of going through an expensive and time-consuming fabrication process, computer simulations can be used to predict the electrical characteristics of a device design quickly and cheaply [14]. Computational prototyping is the main focus of this work that demonstrates how the hierarchical TCAD can be efficiently used for mechanical deformation-aware design and modeling for highly flexible electronic devices.

Figure 8.4 shows a typical schematic view of TFT (a) 2D cut plane of the device (zero strain), (b) under compression, and (c) under tension. When flexed or bent, flexible devices have a specific strain profile, including a neutral line of zero strain. From Figure 8.4, it is clear that when the inner radius of the device is under compression

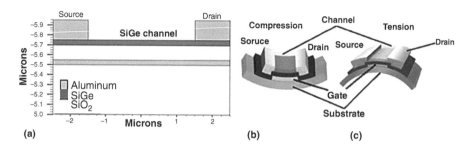

FIGURE 8.4 Schematic view of a TFT, commonly considered in bending test: (a) device, (b) under compression, and (c) under tension.

(b), the outer radius will be under tension (c). At some location in the layer, the stack is the neutral plane (NP) (or neutral line in a 2D cross-sectional presentation) along which there is no tension or compression upon bending. However, the neutral plane exists in a confined volume so it cannot be utilized in flexible devices with multicomponent layered stacked vertically where different types of bending deformations such as tensile, compressive, shearing, bending, and torsional are necessary. In general, reported degradation in electrical performance is based on qualitative stress information [24]. It should be noted that from the perspective of applications, the constraints can be random, and their impact should, as much as possible, be minimized. Unfortunately, there is no systematic method for analyzing the reliability of the flexible electronics that we want to follow at this stage.

In the following, we present a predictive TCAD framework to (a) generate multilayered flexible electronic device structures with large deformations, (b) simulation and analyses stress/strain profiles, and (c) predict the device electrical response under various deformation conditions, which may undergo beyond linear regions. The objectives are to demonstrate the technological feasibility of combining high mechanical flexibility, higher electrical performance, and finally, good stability of the performances under deformation. In the simulation environment, the essential elements of fracture mechanics are briefly recalled first to have a good understanding of the analysis of the strain that takes place in the devices. The modeling of mechanical deformations of multilayered devices is described theoretically. The generation of multilayered SiGe structures with various types of deformations is described next. The impacts of the deformations on output characteristics are estimated via the mobility variation due to the piezoresistive coefficients of SiGe. Microwave characteristics of the deformed transistors have also been analyzed. The simulation results are compared across devices under tensile, compressive, torsional, and zero strain. The effects of bending (radius of curvature) on the device performance have been simulated and compared.

8.1.3 Simulation Environment

Flexible devices are manufactured by stacking multiple thin composite films of different materials, thicknesses, and even sizes. It is essential to look at the mechanical aspects in trying to answer the following question: What will be the impact of external mechanical stresses on the electrical properties of the devices and circuits considered. The flexible term for an electronic component often refers to the deformability of the system. Flexible substrates can be subdivided into two classes: (a) substrates that allow for inextensible bending only and (b) substrates that allow for large arbitrary bending. Any flexible component must be able to withstand deformations of several tens of percent depending on the applications considered [25].

A large deformation takes place by bending, stretching, or twisting flexible devices. Commonly, the flexibility of the devices is studied using analyses based on the neutral plane (or neutral line in a 2D cross-sectional presentation) along which there is no tension or compression upon bending. However, the neutral plane exists in a confined volume. As such, it cannot be utilized in flexible devices with multicomponent vertically stacked layered structures, where different types of bending deformations are involved. To solve issues of mechanical and electrical reliability, the local stress/strain in the device multilayers must be known. Although the mechanical stress of packaged systems on rigid substrates has been studied extensively; however, the mechanical stress distribution in the multilayered flexible structures is rarely discussed in the literature. Most flexible electronic devices are based on inorganic, organic, and printed electronics. The low mobility of these semiconductors and the flexible devices being of the order of a few micrometers limit the components to frequencies of the order of MHz. It is always energetically favorable for the strained system to relax the elastic energy induced by epitaxial strain. Many other mechanisms could be responsible for strain relaxation besides misfit dislocations formation.

The tensile and compressive strain was applied to the TFTs by bending samples to convex and concave curves of known radii. The exact amount of strain in the channel of each TFT is difficult to determine as the samples consist of many layers of different materials patterned to various geometries. However, the strain can be calculated by modeling the bent device. This model treats the various device layers as a single uniform film with Young's modulus, sitting on one surface of the substrate with thickness ds and Young's modulus. For applied tension and compression respectively, the TFTs are represented as either a layer of thickness df1 on the outside of the bend or as a layer of thickness df2 on the inside. The strain on the TFT film surface can then be estimated [26].

A great deal of mechanical analysis in microelectronics relies on the elastic theory. However, solving its fundamental equation is challenging. The solution is usually unknown for varying geometries and conditions faced by the engineering design. Continuum mechanics is the branch of mechanics dealing with the analysis of the kinematics and the mechanical behaviors of materials in terms of strain and stress. For analysis, the finite element method (FEM) can be used to compute the elastic equations on general geometries and conditions. It provides means to simulate elasticity problems by discretizing its fundamental equations. A tensorial treatment will be used, and the body is assumed to be in static equilibrium. Infinitesimal strain theory describes solid behavior for deformations much smaller than the body dimensions [27]. Analyses of small deformations of solids are based on the linearization of

Lagrangian or Eulerian strain tensors of finite strain theory [28], a generalization of infinite strain theory. It was initially envisaged to only deal with bending problems to model the small displacements [29]. In the following, different concepts of mechanics and piezoresistivity used for analyses will be recalled briefly.

In the simulation, we have adopted the reported framework [30] to extend its application in an innovative area, such as strain-engineered multilayered SiGe TFT for flexible electronics applications. When stress is not large, the stress can be calculated using linear elasticity theory (as the stress is proportional to strain) using Young's modulus and Poisson's ratio. One can calculate the stress in the film by applying bending stress at the ends of a beam. However, when the device is formed by stacking several layers of different materials (as in the case of SiGe TFTs) and arbitrary geometry, one needs to consider nonlinear deformation theory to predict the stress.

For an ideal elastic material, the stress–strain relationship can be derived from the strain energy density function, which depends on the deformation gradient matrix [31]. In general, the FEM is used for modeling the mechanical response of isotropic materials. As the flexible devices can experience different types of deformations (tensile, compressive, shearing, bending, and torsional), to deal with any combination of these deformation modes, one can apply prescribed motions at boundaries. The reference frame is made of rectangular coordinate axes at a fixed origin. Applying bending stress to the structure, calculation of stresses caused by rotation and shifting left and right sides of the structure, are performed. Considering rigid surfaces at boundaries such that motion of nodal points at the rigid surface follows the rigid body kinematics, and is given as:

$$x_i = x_{ref} + R\left(\theta_{ref}\right)\left(X_i - X_{ref}\right) \tag{8.17}$$

Where x_{ref} (reference point translation) and θ_{ref} (reference point rotation) are simulation variables, and R is a 3×3 rotation matrix for the given rotation angle. Following reference [30], we consider the linear system of algebraic equations resulting from finite element discretization as:

$$Kd = f \tag{8.18}$$

where d is the unknown displacement vector, f is the given force vector, and K is the symmetric positive definite stiffness matrix. The second Piola–Kirchhoff, S stresses are thus given by:

$$S = DE \tag{8.19}$$

where D are constant elastic moduli and the Green-Lagrange strains, E are given in terms of the deformation tensor as:

$$E = \frac{1}{2}\left(FF^T - I\right) \tag{8.20}$$

where F is the deformation gradient. Cauchy stress, where both the normal and forces are in the deformed frame, is often used in engineering practices. The following transformation can be used to get the Cauchy stresses:

$$\sigma = \frac{1}{\det(F)} F S F^{T} \qquad (8.21)$$

A conjugate strain to the Cauchy stress in large deformations is Almansi strain, which defined as:

$$e = \frac{1}{2}\left(I - F^{-T}F^{-1}\right) \qquad (8.22)$$

In terms of the displacement field, it is written as:

$$e = \frac{1}{2}\left(\frac{\partial u^{T}}{\partial x} + \frac{\partial u}{\partial x} - \frac{\partial u^{T}}{\partial x}\frac{\partial u}{\partial x} \right) \qquad (8.23)$$

The method described above is used to generate different structures of the multi-layered SiGe TFTs on flexible substrates to determine structural integrity, performance, and reliability, as well as predicting the structural failures. The new deformation stress model described above is a very robust simulation method to analyze process-induced deformation stress as well as bending stress, and one can minimize its effect on the device characteristics through simulation without loss of cost and time.

8.2   DEVICE STRUCTURE GENERATION

As the first step for the structure's deformation-induced stress analysis, several structures are generated. Attempts have been made to determine of stress profile in bent and twisted structures. Calculation of stress caused by such deformation has also been performed. To this end, various end rotation angles (tensile and compressive) are applied, and results are summarized in Figure 8.5. Figure 8.6 shows the device dimensions with the torsional (sine) bending. The radius of curvature is 2 μm. The stress profile (in XX-direction) is also shown.

VictoryMesh [32] is used to create SiGe TFT planar device structures with multi-layered materials and thicknesses. The layer thickness can be varied in the vertical direction. In the planar device, three layers are specified in the x-axis and five layers in the z-axis on which the electrodes are created (Figure 8.5a). The planar device structure will then be used for various types of bending applications. The origin of the bending structure is set such that the bend is compressive (parallel, Figure 8.5b) and tensile (parallel, Figure 8.5c) with the channel when the polar bend is generated in the x-axis. When the sinusoid transform mode of the bend is applied in both the horizontal axes, a sinusoidal (sine) bending (Figure 8.5d) takes place in the structure. In a typical application, there may be multiple forms of bending impacting the device performance at the same time.

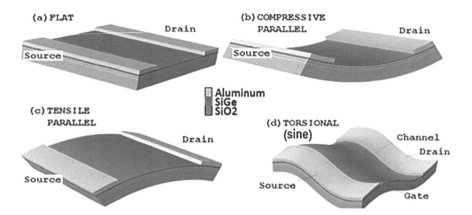

FIGURE 8.5 Different bending configurations generated from the planar TFT structure: (a) Flat (no stress), (b) compressive bending (parallel to the current direction, CP), (c) tensile bending (parallel to the current direction, TP), (d) torsional (sine) bending. The radius of curvature, RC = 2 μm for both CP and TP.

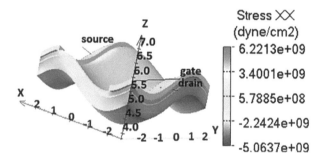

FIGURE 8.6 Typical stress profile (in *XX*-direction) in torsional (sine) bending. The radius of curvature is 2 μm. Device dimensions are also shown.

8.3 STRESS ANALYSIS

This simulation experiment aimed to investigate the influence of the bending geometry (curvature) on device characteristics. To investigate mechanical bending effects on electrical characteristics, we perform several 3D TCAD simulations for different uniaxial bending radii along channel length for the structure (Figure 8.5b) compressive bending (parallel to current direction). If some external or internal forces are applied to the body, it deforms or changes its size. Even if the body is deformed, the body represents the continuum before and after deformation. Therefore, the coordinates of all body points after deformation are continuous, specific functions of corresponding initial coordinates. The elastic model solves the linearized elastic equations. It provides the displacement vectors in device and crystal coordinates. From the displacement vectors, the strain tensor is computed on the mesh elements.

The elastic model takes strain due to a lattice mismatch between materials into account. The lattice constants are retrieved from the material database. In the simulation, an initial strain condition is applied to all segments forcing their lattice constant to match that of the substrate segments. Solving the elastic equations allows the system to relax and minimize its energy as a whole, thus obtaining the correct strain conditions at material interfaces. The elasticity tensor defines the relationship between stress and strain with the components C_{11}, C_{12}, and C_{44}. The components and the lattice constants are retrieved from the material database.

For generating different curvatures to the planar device, we bend the structure to rotate left and right sides by 30 degrees clockwise and anticlockwise around the geometrical center, correspondingly. Then we export the structure with deformation information in the structure for visualization purposes. Figure 8.7 shows the two cases of bending for compressive and tensile stresses with a radius of 2 μm radius. The relationship between device dimension and stress distribution is found to depend on the material's Young's moduli, for example, the larger the areas of the material in the device, the less stress during mechanical strain. The cut plane may be used to

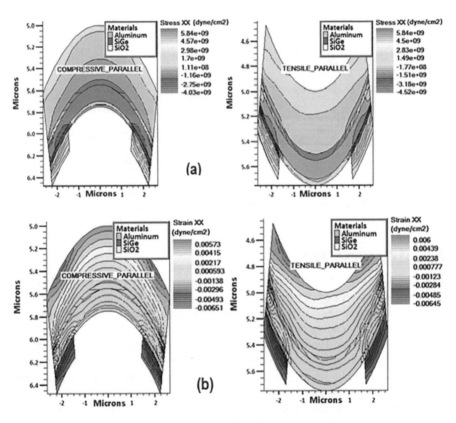

FIGURE 8.7 2D cut plane (XZ) at $y = 0$ with curved compressive and tensile structures generated (30-degree end rotation) with a bending radius of 2 μm. Both the XX-stress (a) and XX-strain contours (b) are shown.

examine the inside of a structure or to perform a 2D device simulation. Both the XX-stress and XX-strain contours are shown. The simulated stress shows lower stress as the bending radius increases. These components of the strains and stresses are, for reasons of symmetry, only plotted against the thickness of the deformed flexible structure.

VictoryStress [33] is used for the stress/strain calculations, and the command "rebalance stress" was used, which allows for free movement of side walls in the structure. After the "STRESS" statement is processed, the structure contains the displacement field describing the structure's deformation caused by the existing stresses, as described in the simulation environment. After the computation of stress/stress, the structure contains the displacement field describing the structure's deformation (shape) caused by the existing stresses. The displacement profiles are plotted. A cut plane is a plane (2D slice) that is drawn through a 3D structure. The cut plane may be used so one can examine the inside of a structure or to perform a 2D device simulation. We have calculated the internal stresses caused by the deformation of the flexible structures.

The von Mises stress along the interface is essential for the study of film reliability because the shear stress and the tensile stress contribute to the driving force for interfacial delamination. Von Mises stress is a single indicator combining shear and tensile stresses. In the case of uniaxial stress, state stain energy density is equal to the area under the stress–strain curve. For the general 3D case, the strain energy density, U_0 is expressed as:

$$U_0 = \frac{1}{2}\sigma_x \varepsilon_x + \sigma_y \varepsilon_y + \sigma_z \varepsilon_z + \tau_{yz}\gamma_{yz} + \tau_{zx}\gamma_{zx} + \tau_{xy}\gamma_{xy} \tag{8.24}$$

where σ_x, σ_y, and σ_z are normal stress components, ε_x, ε_y, and ε_z are strain components, τ_{yz}, τ_{zx}, and τ_{xy} are shear stress components, γ_{yz}, γ_{zx}, and γ_{xy} are surface energy per unit area.

For a coordinate system that is parallel to the principal stress directions, no shear components exist. Extending Equation (8.24) to this stress states yields:

$$U_0 = \frac{1}{2}\left(\sigma_1 \varepsilon_1 + \sigma_2 \varepsilon_2 + \sigma_3 \varepsilon_3\right) \tag{8.25}$$

where σ_1, σ_2, and σ_3 are normal stress components and ε_1, ε_2, and ε_3 are strain components. As stresses and strains are related through the linear elastic relations, principal stresses and strains are given as

$$\begin{cases} \varepsilon_1 = \dfrac{1}{E}\left(\sigma_1 - v\sigma_2 - v\sigma_3\right) \\[2mm] \varepsilon_2 = \dfrac{1}{E}\left(\sigma_2 - v\sigma_1 - v\sigma_3\right) \\[2mm] \varepsilon_3 = \dfrac{1}{E}\left(\sigma_3 - v\sigma_1 - v\sigma_2\right) \end{cases} \tag{8.26}$$

where v is Poisson's ratio, and E is Young's modulus.

Substituting from Equation (8.25) into Equation (8.26), the strain energy density in terms of principal stresses is given as

$$U_0 = \frac{1}{2E}\left[\sigma_1^2 + \sigma_2^2 + \sigma_3^2 - 2v\left(\sigma_1\sigma_2 + \sigma_2\sigma_3 + \sigma_1\sigma_3\right)\right] \qquad (8.27)$$

The von Mises stress σ_{VM} is defined in terms of principal stresses as

$$\sigma_{VM} = \sqrt{\frac{\left(\sigma_1 - \sigma_2\right)^2 + \left(\sigma_2 - \sigma_3\right)^2 + \left(\sigma_3 - \sigma_1\right)^2}{2}} \qquad (8.28)$$

The von Mises stress can be rewritten in terms of stress components as

$$\sigma_{VM} = \sqrt{\frac{\left(\sigma_{xx} - \sigma_{yy}\right)^2 + \left(\sigma_{yy} - \sigma_{zz}\right)^2 + \left(\sigma_{zz} - \sigma_{xx}\right)^2 + 6\left(\tau_{xy}^2 + \tau_{yz}^2 + \tau_{zx}^2\right)}{2}} \qquad (8.29)$$

where σ_{xx}, σ_{yy}, and σ_{zz} are normal stress components.

Figure 8.8 shows the 2D von Mises stress (XZ) distribution in compressively and tensilely bent TFT structure at a bending radius of 2 µm. It is observed that decreasing bending radius induces considerable strain in TFTs. Also, we can see, most of the stress appears at the metal edge of the source/drain contact material nearby the bending point.

Figure 8.9 shows the 2D von Mises stress (XZ) distribution in compressively and tensilely bent TFT structures with different Ge content at a bending radius of 2 µm. The simulated stress shows the lowest stress for a radius of 2 µm. It is observed that decreasing bending radius induces considerable strain in TFTs. Also, as Ge content increases von Mises stress increases, and most of the stress appears at the metal edge of source/drain contact material nearby the bending point.

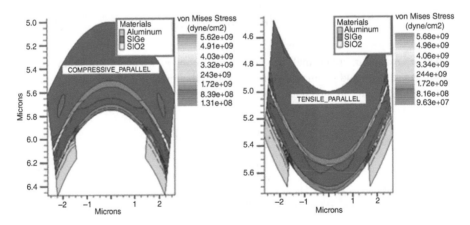

FIGURE 8.8 2D von Mises stress (XZ) distribution in compressively and tensilely bent TFT structure at a bending radius of 2 µm.

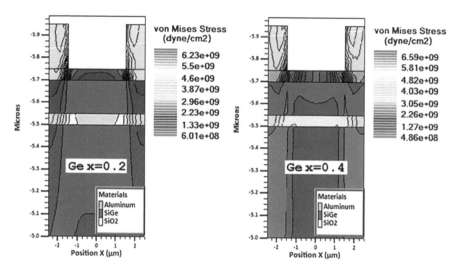

FIGURE 8.9   2D von Mises stress (XZ) distribution in compressively and tensilely bent TFT structures at a bending radius of 2 μm as a function of Ge content. The simulated stress decreases as the bending radius increases.

8.3.1 THERMAL EXPANSION

SiGe alloy is a promising candidate for thermoelectric materials; while it shows a significantly reduced thermal conductivity (κ) as compared to pure Si and Ge. The lattice thermal conductivity for SiGe decreases significantly with increasing germanium content. The thermal conductivity of silicon and germanium nanostructures substantially differs from those of bulk materials due to the size effect and high surface to volume ratio. The reduction of thermal conductivity is a critical issue as junction temperature rises due to a reduction in heat dissipation. One of the sources of strain/stress is the thermal expansion mismatch between material layers (e.g., between the substrate and overlying films). This source of strain is external and is given by [34]

$$\varepsilon_{ii}^0 = \int_{T_1}^{T_2} \alpha(T)\, dT \tag{8.30}$$

where $\alpha(T)$ is the coefficient of thermal linear expansion, and temperature changes from $T1$ to $T2$ Kelvin. The thermal linear expansion coefficient is a function of absolute temperature in the Kelvin scale. If this coefficient is a constant then Equation (8.30) will be as follows:

$$\varepsilon_{ii}^0 = \alpha\, \Delta T = \alpha\left(T_2 - T_1\right) \tag{8.31}$$

Note that the volume of material changes in each normal direction. Therefore, all initial shear strain components related to this thermal lattice mismatch will be equal to zero. A comparison of 2D von Mises stress (XZ) distribution due to thermal stress

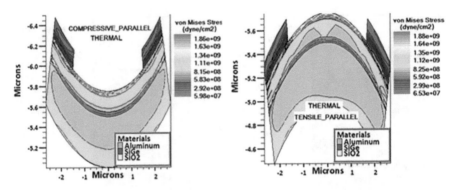

FIGURE 8.10   2D von Mises thermal stress (XZ) distribution comparison in compressively (a) and tensilely (b) strained SiGe TFT structures at a bending radius of 2 µm. The simulated von Mises stress is found to be lower in compressively strained structures.

in compressively and tensilely strained SiGe TFT structures is shown in Figure 8.10. Compressively bent structure is found to have a lower von Mises thermal stress.

8.4   DEVICE SIMULATION

The use of numerical simulation has become more important in the semiconductor industry because it is a tool that has proven to be able to provide an accurate prediction for the development of future devices. Considering the diverse requirements in device simulation, it is natural to ask for the level of confidence at which device simulation tools are useful for device development. For accurate prediction, the process and device simulators in the TCAD suite of tools must be calibrated [35]. To be "predictive" TCAD needs to have a very high level of accuracy. However, properly required calibration for predictive TCAD is time-consuming and expensive to calibrate the simulators. The device simulation calibration focuses first on tuning the transport models. The device electrostatics are adjusted using the low and high drain voltage biased I_d-V_g and I_d-V_d characteristics. In the simulation, the low field mobility tuning is primarily based on phonon scattering and impurity scattering mobility models. The high field mobility tuning adjusts the saturation velocity and the critical field in the field-dependent mobility models.

Due to the deformation in the channel length resulting from bending, variations in the electrical performances are expected. They are compared against the planar device using the device simulator VictoryDevice [36]. The key material properties and device geometry parameters used in the simulation are shown in Table 8.1. In stress simulations, Young's modulus and Poisson's ratio were used to predicting the optimum material stack thickness and size through comprehensive numerical simulation. Young's modulus (dyne/cm²) and Poisson's ratio values for SiGe, aluminum, and SiO₂ are also shown in Table 8.1.

The impact of the deformations on the electrical properties is estimated via the variation of mobility due to the piezoresistive coefficients. The electric field and potential profile in flat (no stress condition) and torsional bending conditions are

TABLE 8.1

Key Material Properties and Device Geometry Parameters Used in the Simulation

Device Parameters	Values
TFT Gate Length (L)	3 μm
TFT Width (W)	5 μm
SiO$_2$ Thickness	0.15 μm
Substrate (SiO$_2$) Thickness	0.6 μm
SiGe Layer Thickness	0.05 μm
Al Contact Thickness	0.2 μm
SiGe Young's Modulus (dyne/cm^2)	1.31e12
SiGe Poisson's Ratio	0.36
SiO$_2$ Youngs Modulus (dyne/cm^2)	0.7e12
SiO$_2$ Poisson's Ratio	0.17
Aluminum Young's Modulus (dyne/cm^2)	7e11
Aluminum Poisson's Ratio	0.35

shown in Figure 8.11. The static figures of merit (I_d-V_g and I_d-V_d) drain current, static transconductance, and microwave (cutoff frequency, f_T, and maximum oscillation frequency, f_{max}) is considered proportional to the mobility of majority carriers. The variations in stresses translate directly into a variation in drain current, static transconductance, and characteristic frequencies f_T and f_{max}. The I_d-V_g and I_d-V_d characteristics for the 2 μm radius of curvature are shown in Figures 8.12 and 8.13, respectively.

In general, the electrical characteristics of SiGe TFTs are degraded with bending strain, reducing the field-effect mobility, and on current. A possible reason for the variation comes from the modification in the localized DOS in the bandgap. The bending strain generates more broken/weak bonds inside the SiGe channel region, increasing densities of defect states. With increasing stress, the defect states with increased densities capture or scatter more electrons, lowering the current and field-effect mobility.

As expected, as the bending is changed from compressive (parallel to the current direction) to tensile (parallel to the current direction), the drain current is found to decrease. We believe that current degradation takes place due to the change in SiGe material with different stress conditions in the multilayered structure. We believe that the reason why the compressively stressed device shows a larger current than the tensile case is related to the reduced effective channel length shown by the majority carrier at the front channel interface. Several other essential device parameters such as V_T, subthreshold slope (SS), I_{on}, I_{off}, Theta (mobility degradation factor), and Beta (transconductance parameter) can be extracted from the I_d-V_g characteristics for all the cases considered.

The single most important parameter used to characterize the high-frequency behavior of a transistor is the common emitter current gain-bandwidth product f_T, also called the transition frequency. Another very important parameter that characterizes the high-frequency operation of the RF transistors is the maximum oscillation

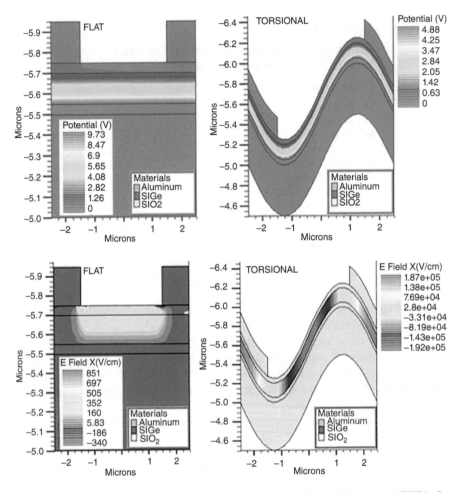

FIGURE 8.11 Comparison of electric field and potential profiles in SiGe channel TFT in flat (no stress condition) and torsional bending conditions.

frequency or unity power gain frequency, f_max. F_{max} is often more important than f_T and is a good measure of transistor performance not only for power gain in small-signal and large-signal amplifiers but also for wideband analog amplifiers and even for non-saturating logic gates.

Maximum available gain (MAG) is obtained when both input and output are simultaneously conjugately matched. MAG exists only when the device is unconditionally stable when $k > 1$. U equals maximum available unilateral gain (MAUG) only if the device is unilateral, that is, $y_{12} = 0$. MAG and MSG are equal to each other once the device is unconditionally stable. The frequency at which MAG becomes unity is often defined as f_{max}.

The cutoff frequency is normally extracted for various operating points. This so-called unity current gain frequency can be calculated by extrapolation because a conventional transistor drops with a slope of dB/dec or dB/octave at higher

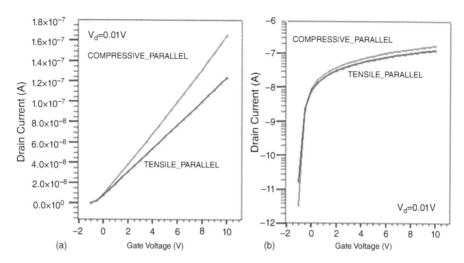

FIGURE 8.12 Transfer characteristics of SiGe TFTs for 2 μm radius of curvature (compressive and tensile) at a $V_d = 0.1$ V. The compressively strained structure shows higher current due to higher stress generation. A small radius of curvature (more bending) will generate higher stress and subsequently, a higher drain current.

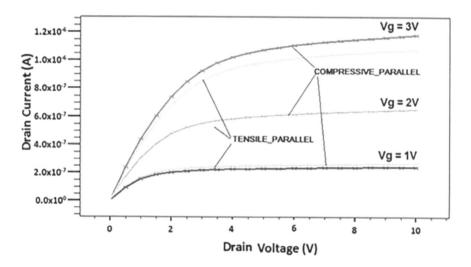

FIGURE 8.13 Drain current comparison of stressed (compressive and tensile) devices with the radius of curvatures: 2 μm for gate voltage $V_g = 3$ V. Compressively stressed 2 μm curvature devices show highest drain current compared tensilely strained devices due to higher strain in the channel.

frequencies. Thus, the frequency is increased until the dB/dec region of the curve is reached. An extrapolation yields the cutoff frequency. Measurement equipment often uses this approach to obtain f_T, since they are normally not able to measure such high frequencies as required for today's cutoff frequencies. However, this method depends on the frequency chosen that means which frequency assures the validity of the single-pole approximation. For simulation purposes, it is very inconvenient to run a simulator and a post-processing script in the end. For that reason, a conditional stepping approach was developed to use a mathematical iteration algorithm to approximate for a given accuracy.

The second important RF figure of merit is the maximum oscillation frequency f_{max}, which is related to the frequency at which the device power gain equals unity. The value can be determined in two ways. The first one is based on the unilateral power gain as defined by Mason. Therefore, f_{max} is the maximum frequency at which the transistor still provides a power gain. An ideal oscillator would still be expected to operate at this frequency, and hence the name maximum oscillation frequency. The second way to determine which not entirely correct is based on the MAG and the maximum stable gain (MSG). Whereas MAG shows no definite slope, MSG drops with a 20 dB/dec slope. Generally, transistors have useful power gains up to that above they cannot be used as power amplifiers anymore. However, the importance and depends on the specific application. Thus, there is no general answer whether should be prioritized over. Both figures should be as high as possible and manufacturers often strive to enter many different markets for their transistors.

There are several ways to predict the small-signal and large-signal high-frequency properties of semiconductor devices. A review of these different techniques has been given by Laux et al. [32]. Frequency domain perturbation analysis is used to calculate the small-signal characteristics, while Fourier analysis is required for a large-signal response. In ATLAS, frequency domain perturbation of a dc solution can be used to calculate small-signal characteristics at any frequency. These y-parameters can then be used to find different power gains [33]. Among the various power gains described so far in the literature several, such as MAG, MSG, and MAUG, have found widespread use. Additionally, a figure-of-merit that has been used extensively for microwave characterization is Mason's invariant U (or Mason's gain). These quantities are calculated from the measured small-signal scattering parameters because of the ease of measurement at high frequencies.

The AC performance of the flexible devices was evaluated using TCAD simulation of two-port scattering parameters. The microwave characteristics of the deformed transistors were analyzed using the unity gain power frequency from the MSG and the unilateral gain. These parameters are calculated from the scattering parameters using Equations (8.32)–(8.34) [37]:

$$|H_{21}| = \frac{2S_{21}}{(1 - S_{11})(1 + S_{22}) + S_{12}S_{21}} \tag{8.32}$$

$$U = \frac{\left|\frac{S_{21}}{S_{12}} - 1\right|^2}{2\left[k\left|\frac{S_{21}}{S_{12}}\right| - \Re\left(\frac{S_{21}}{S_{12}}\right)\right]} \tag{8.33}$$

$$k = \frac{1-\left|S_{11}\right|^2 - \left|S_{22}\right|^2 + \left|S_{11}S_{22} - S_{12}S_{21}\right|^2}{2\left|S_{12}\right|\left|S_{21}\right|} \qquad (8.34)$$

where S_{11}, S_{12}, S_{22}, and S_{21} are the frequency-dependent scattering parameters whereas port 1 is connected to the gate, and port 2 is connected to the drain.

Figure 8.14 shows the unilateral power for the 2 μm radius of curvature considered. The current gain H_{21} and of Mason U are presented with configurations for bending radius 2 μm. It is important to note that the curve follows the theoretical slope of −20 dB/dec allowing reliable extraction of the characteristic frequencies f_T and f_{max}. Besides, in contrast to static dc performance, very little difference seems visible between different levels of strain. The influence of bending on flexible SiGe TFTs, exhibiting a maximum oscillation frequency (maximum power gain frequency) f_{max} beyond 300 MHz has been reported [37]. Interestingly, the simulation results on the microwave characteristics show possible higher performance ($f_T > 700$ MHz) compared to the NP engineered a-IGZO TFTs ($f_{max} > 300$ MHz). This high frequency is achieved due to the short channel length of ~1 μm in combination with a low gate capacitance.

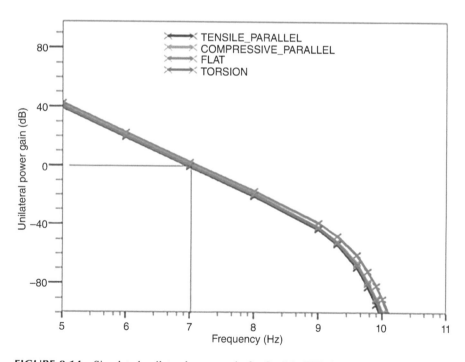

FIGURE 8.14 Simulated unilateral power gain for flexible TFTs bent with torsional, compressive, and tensile curvature of radius 2 μm, respectively is shown. For comparison purposes, no strain condition cutoff frequency is also shown. A cutoff frequency, f_T of ~690 MHz is observed in all cases.

8.4.1 SUMMARY

Following More-than-Moore applications, in this chapter, we have presented the design and development of a strain-engineered flexible SiGe channel TFT. A brief review of the growth techniques of the strained-SiGe, strained-Si films, and the relaxed SiGe virtual substrates is presented. Chemical vapor deposition is the preferred technique for epitaxial film growth in production and industrial environments. Strain relaxed SiGe substrates can be grown on silicon with thickness well above the critical thickness. To minimize the dislocation density in these layers, the Ge composition is graded from 0% up to the required value with a slow grading rate below 10% Ge per μm at high growth temperatures (~800°C). The electronic properties of $Si_{1-x}Ge_x$ and strained-Si have been discussed.

SiGe film-based flexible TFTs can be produced by altered silicon CMOS processes, which facilitates the adoption of the technology by the industry as there is no need for investment in novel fabrication equipment. Existing knowledge of a-Si:H TFT fabrication can be applied toward the development of flexible SiGe TFTs as they are based on relatively mature technology.

We have demonstrated a simple and viable approach to realizing strained-SiGe RF transistors on flexible plastic substrates. This technique has great potential in low-power and high-speed flexible-electronics applications and could be used to replace some rigid counterparts for use in mechanically bendable and non-planar conformal surfaces where rigid devices cannot be easily used. One can foresee as a consequence manufacturable large-area applications of such flexible high-speed TFT technology.

Recent efforts have been focused on functionality tuning as well as potential applications in electronics by taking advantage of the vertical strain, defect, and interface. However, there are still many open questions that include, but are not limited to vertical strain is certainly one of the most critical factors in tuning a variety of physical properties. However, the role of defects and unique microstructure on the physical properties such as electron transport, ionic conduction, superconductivity, and so on needs to be fully investigated.

REFERENCES

[1] IEEE, "International roadmap for devices and systems, 2020 update, beyond CMOS." 2020.

[2] IEEE, "International roadmap for devices and systems, 2020 update, more Moore." 2020.

[3] M. G. Kane, "AMOLED display technology and applications," *Flexible Carbon-based Electronics*. John Wiley & Sons Ltd, pp. 231–263, 2018, doi: 10.1002/9783527804894. ch8.

[4] M. Nayfeh, Ed., "Chapter 14 - electronics and communication," *Fundamentals and Applications of Nano Silicon in Plasmonics and Fullerenes*. Elsevier, pp. 431–485, 2018, doi: 10.1016/B978-0-323-48057-4.00014-1.

[5] P. Heremans, "*Electronics on plastic foil, for applications in flexible OLED displays, sensor arrays and circuits,*" in *2014 21st International Workshop on Active-Matrix Flatpanel Displays and Devices (AM-FPD)*, 2014, pp. 1–4, doi: 10.1109/AM-FPD.2014.6867106.

[6] H. Oh, K. Cho, and S. Kim, "Electrical characteristics of a bendable a-Si:H thin film transistor with overlapped gate and source/drain regions," *Appl. Phys. Lett.*, vol. 110, no. 9, 2017, doi: 10.1063/1.4977564.

[7] T. Ma, V. Moroz, R. Borges, K. E. Sayed, P. Asenov, and A. Asenov, *"(Invited) Future perspectives of TCAD in the industry,"* in *2016 International Conference on Simulation of Semiconductor Processes and Devices (SISPAD)*, 2016, pp. 335–339, doi: 10.1109/SISPAD.2016.7605215.

[8] I. A. Lysenko, L. A. Patrashanu, and D. D. Zykov, *"Organic light emitting diode simulation using Silvaco TCAD tools,"* in *2016 International Siberian Conference on Control and Communications (SIBCON)*, 2016, pp. 1–5, doi: 10.1109/SIBCON.2016.7491782.

[9] S. A. Scott and M. G. Lagally, "Elastically strain-sharing nanomembranes: Flexible and transferable strained silicon and silicon–germanium alloys," *J. Phys. D. Appl. Phys.*, vol. 40, no. 4, pp. R75–R92, Feb. 2007, doi: 10.1088/0022-3727/40/4/R01.

[10] C. K. Maiti, S. Chattopadhyay, and L. K. Bera, *Strained-Si Heterostructure Field Effect Devices*. CRC Press, Boca Raton, 2007.

[11] A. Schulze et al., "Observation and understanding of anisotropic strain relaxation in selectively grown SiGe fin structures," *Nanotechnology*, vol. 28, no. 14, 2017, doi: 10.1088/1361-6528/aa5fbb.

[12] Y. M. Haddara and M. M. R. Howlader, "Integration of heterogeneous materials for wearable sensors," *Polymers (Basel)*, vol. 10, no. 1, 2018, doi: 10.3390/polym10010060.

[13] A. Nainani et al., *"Is strain engineering scalable in FinFET era?: Teaching the old dog some new tricks,"* in *2012 International Electron Devices Meeting*, 2012, pp. 18.3.1–18.3.4, doi: 10.1109/IEDM.2012.6479065.

[14] G. A. Armstrong and C. K. Maiti, *TCAD for Si, SiGe, and GaAs Integrated Circuits*. The Institution of Engineering and Technology (IET), UK, 2008.

[15] S. Das, "Design and simulation of bandgap-engineered MOSFETs," PhD Thesis, SOA University, 2019.

[16] F. C. Frank and J. H. Van der Merwe, "One-dimensional dislocations. II. Misfitting monolayers and oriented overgrowth," *Proc. Roy. Soc.*, vol. A198, pp. 216–225, 1949.

[17] E. Kasper and E. K. Lyutovich, *Properties of Silicon Germanium and SiGe:Carbon*. IEE-INSPEC, London, 2000.

[18] D. J. Paul, "Silicon-germanium strained layer materials in microelectronics," *Adv. Mater.*, vol. 11, pp. 191–204, 1999.

[19] C. K. Maiti and G. A. Armstrong, *Applications of Silicon-Germanium Heterostructure Devices*. Inst. of Physics Publishing, Bristol, 2001.

[20] S. E. Thompson, G. Sun, K. Wu, J. Kim, and T. Nishida, *"Key differences for process-induced uniaxial vs. substrate-induced biaxial stressed Si and Ge channel MOSFETs,"* in *IEEE IEDM Tech. Dig.*, 2004, pp. 221–224.

[21] C. K. Maiti, N. B. Chakrabarti, and S. K. Ray, *Strained Silicon Heterostructures: Materials and Devices*, The IEE, UK, 2001.

[22] D. J. Paul, "Si/SiGe heterostructures: From material and physics to devices and circuits," *Semicond. Sci. Technol.*, vol. 19, pp. R75–R108, 2004.

[23] T. Vogelsang and K. R. Hofmann, "Electron transport in strained Si layers on $Si_{1-x}Ge_x$ substrates," *Appl. Phys. Lett.*, vol. 63, pp. 186–188, 1993.

[24] B. Chen et al., "Effects of repetitive mechanical bending strain on various dimensions of foldable low temperature polysilicon TFTs fabricated on polyimide," *IEEE Electron Device Lett.*, vol. 37, no. 8, pp. 1010–1013, 2016, doi: 10.1109/LED.2016.2584138.

[25] S. Huang, Y. Liu, Y. Zhao, Z. Ren, and C. F. Guo, "Flexible electronics: Stretchable electrodes and their future," *Adv. Funct. Mater.*, vol. 29, no. 6, p. 1805924, 2019, doi: 10.1002/adfm.201805924.

[26] H. Gleskova, S. Wagner, and Z. Suo, "Failure resistance of amorphous silicon transistors under extreme in-plane strain," *Appl. Phys. Lett.*, vol. 75, no. 19, pp. 3011–3013, 1999, doi: 10.1063/1.125174.

[27] G. Q. Zhang, W. van Driel, and X. Fan, *Mechanics of Microelectronics*. Springer, 2006.

[28] de Souza Neto D. Pericand D. R. J. Owen, *Computational Methods for Plasticity: Theory and Applications*. John Wiley & Sons, 2011.

[29] M. M. Billah, M. M. Hasan, M. D. H. Chowdhury, and J. Jang, "P-10: Excellent mechanical bending stability of flexible a-IGZO TFT by dual gate dual sweep using TCAD simulation," *SID Symp. Dig. Tech. Pap.*, vol. 47, no. 1, pp. 1155–1158, 2016, doi: 10.1002/sdtp.10842.

[30] H. Lim, S. Kong, E. Guichard, and A. Hoessinger, "*A general approach for deformation induced stress on flexible electronics,*" in *2018 International Conference on Simulation of Semiconductor Processes and Devices (SISPAD)*, 2018, pp. 276–279, doi: 10.1109/SISPAD.2018.8551752.

[31] P. G. Ciarlet, *Mathematical Elasticity, Vol. 1, Three-Dimensional Elasticity*. Elsevier Science Publishers B. V., 1988.

[32] Silvaco International, VictoryMesh user manual, 2018.

[33] Silvaco International, VictoryStress user manual, 2018.

[34] Silvaco International, VictoryProcess user manual, 2018.

[35] C. K. Maiti, *Introducing Technology Computer-Aided Design (TCAD) Fundamentals, Simulations, and Applications*. CRC Press (Taylor and Francis), Pan Stanford, USA, 2017.

[36] Silvaco International, VictoryDevice user manual, 2018.

[37] N. Munzenrieder et al., "Flexible InGaZnO TFTs with f_{max} above 300 MHz," *IEEE Electron Device Lett.*, vol. 39, no. 9, pp. 1310–1313, 2018.

Index